W9-DGN-487

Fish On, Fish Off

Fish On, Fish Off

The Misadventures and Odd Encounters of the Self-Taught Angler

Stephen Sautner

Guilford, Connecticut

An imprint of Rowman & Littlefield

Distributed by NATIONAL BOOK NETWORK

Copyright © 2016 by Stephen Sautner

All rights reserved. No part of this book may be reproduced in any form or by any electronic or mechanical means, including information storage and retrieval systems, without written permission from the publisher, except by a reviewer who may quote passages in a review.

British Library Cataloguing in Publication Information Available

Library of Congress Cataloging-in-Publication Data available

ISBN 978-1-4930-2505-3 (hardcover)
ISBN 978-1-4930-2506-0 (electronic)

∞™ The paper used in this publication meets the minimum requirements of American National Standard for Information Sciences—Permanence of Paper for Printed Library Materials, ANSI/NISO Z39.48-1992.

Then terrible things happened, there in the murky water and the misty moonshine. A monster of the deep tried to do me to death—that's what happened. To this day I carry the physical scars of that encounter; and there's a scar on my mind too, a mute testimonial of my ordeal on that dread night.

—Archibald Rutledge, *"Death in the Moonlight"*

Dedicated to the memory of Dery Bennett (1930–2009).
Friend. Conservationist. Mischief Maker.

Contents

Foreword

Fish On, Fish Off is a well-guided fishing trip. It is also a tonic and antidote.

I have spent decades reading and editing copy about, say, brick-heavy brook trout the color of Canadian sunsets that always slide into landing nets just as barbless flies slip from their jaws. And there have been way too many silver bonefish that streak across golden flats and are invariably brought to hand with no mention of such realities as: the strands of white slime dangling from the leader; the broken rod tip that preceded a sloppy cast; or the guide whose command of English begins and ends with "Shut up."

Poetry in any kind of writing is great, provided it's good poetry that allows the reader to occasionally come up for air. Finding that quality and balance in hook-and-bullet writing is like finding guilelessness in Congress. But you will find it here.

Along with good poetry sparingly applied, *Fish On, Fish Off* provides us with something we've longed for even more—unvarnished truth presented with wit and humor. Sautner shows us that some of the best and funniest adventures are "misadventures." As he accurately notes, we "relish these tales—sometimes even more than the traditional big fish story."

Like all good writers Sautner has a talent for beginning a piece. But he also excels at the more important and far more difficult task of *ending* a piece. Examples: "Maybe the Falklands mullet [not a mullet at all, but a fine and tasty gamefish] is best where it belongs—on its namesake remote outpost in the lonely South Atlantic." And: "All I could do was look at the floating tree, its salt-bleached limbs bobbing back and forth in the boat's wake as if waving a long, sad goodbye."

This angler's angler has done it all from tigerfish in Zambia to sail-fish in Guatemala to chubs feeding on bird poop under a rusty New York bridge. Because he doesn't rank species or specialize in any, his angling opportunities are endless. As he accurately notes on these pages, "we are proud practitioners of the high art of jigging for panfish."

I knew what to expect from *Fish On, Fish Off* because I'd read Sautner's outdoor columns in the *New York Times* and because on February 17, 2002, he and his wife, Mimi, met me on a woodland pond in western Massachusetts. That expedition happened before the advent of the "Finnish auger with surgically sharpened blades" that Sautner writes about here (and that recently sliced the light switch off my office wall when it slipped to the floor). So, under croaking ravens and rain mixing with snow, we exercised our arms and hearts by chopping holes in thick ice with a spud made by my machinist friend who knows how to harden steel and weld it. What the perch lacked in quantity they more than made up for in size—plump, crimson-finned cows of twelve to fourteen inches. Steve couldn't have been more thrilled if we'd all doubled on Atlantic salmon.

That's because he understands and teaches that fishing is not a mere "sport," like bowling, but immersion in and participation with nature. He pays attention to and appreciates what's going on around him, a talent that elevates his prose and serves him well in his work as executive director of communications for the Wildlife Conservation Society.

What other outdoor writers might describe as "birdsong," if they noticed it at all, Sautner describes as "metallic rattles of veeries and sad, fluty notes of hermit thrushes." The sea that nearly killed him isn't just "angry," it is a "cold, uncaring witch" and he doesn't just get "flattened" by one of its towering waves, he becomes "a flea . . . swatted by the hand of Poseidon himself." The unexpected strike of a big striper in high surf in the final seconds of a fruitless night isn't a "welcome hit," not even a "longed-for jolt." It is a "Gatorade bath after a come-from-behind touchdown as time expires."

You don't just read about what Sautner does; you're on scene, doing it with him. Join him now in this fishing *tour de force.*

—Ted Williams, conservation editor, Fly Rod & Reel magazine

Introduction: Fish On

Ever since the first caveman carved a hook out of a bone and baited it with a hunk of mammoth meat, anglers have been bragging about some big-ass fish they've caught. Just look at cave drawings and petroglyphs of tuna, or salmon, or sharks. What are these but Neolithic boasts of angling glory?

Nowadays, hang around any tackle shop, dock, bar, or Internet forum and you'll hear this tradition carried on (and on). Most follow this simple narrative: "There I was . . . Suddenly, there it was! Fish on! The fish of a lifetime! Somehow—and against all odds—I managed to land the monster! I'm very modest about the whole thing—by the way did I show you where it sits above my mantel? Oh and here's the shadow box where I retired the fly. And did I mention I've named my daughter Four-Pound Brookie?"

Unless you are the second coming of Zane Grey, I say enough already.

Fortunately, there is a yin to this yang of endless angling heroics. Without it, fishing would be inherently dull and one-sided. I am referring to the great untold fish stories: angling experiences that are steeped in humiliation, or pain, or fear. Like the time when the canoe flipped as you were trying to unsnag a lure from a tree branch. Or when you went night fishing for the first time, and two raccoons decided to mate—loudly—in the woods twenty feet behind you. Or maybe you never even made a cast because the wind blew your car door shut with the keys—not to mention your tackle—locked inside.

Some of us relish these tales—sometimes even more than the traditional big fish story. We are the same ones who watched the opening montage of ABC's *Wide World of Sports* in the 1970s just to see the "agony of defeat" scene—when the ski jumper spectacularly crashed off the slope. As soon as he was done careening through the crowd, we turned the channel.

But anglers can be reluctant to regale their fellow anglers with stories of failure. You don't see too many of us walk into the local tackle shop, puff out our chest, and declare: "Y'all gather 'round and I'll tell you about the time I shattered my rod in the car door." For that matter, where's the cave drawing of Gak, the mighty angler, getting chased from his favorite fishing hole by a mastodon?

Anyone who has fished long enough will have logged at least a few misadventures. The ones you are about to read are mine. They are not the types of fishing stories I wrote for the *New York Times*'s "Outdoors" column. There I attempted—at least after longtime editor Susan B. Adams got through with them—to show the poetry of angling. These stories show the scars.

Coming from a non-fishing family, self-taught, and as they used to say in eighth grade "a spaz," maybe I have more stories of angling missteps than most, but I don't think so. Instead, I believe that my fellow anglers are holding back—perhaps the only time when fishermen show some modesty.

For me, some fishing-trips-gone-wrong have been among my most memorable, still seared into memory banks decades after they happened. Maybe it's the brain's way of warning me not to do this again. This holds true whether it was a pond I pedaled my bike to in high school, or some far flung fishing ground half a world away. Yet not all of them were awful. Indeed, some of my finest angling moments rose from the ashes of what began as an otherwise horrible day.

Charles Fox, who probably caught more big trout than anyone on the Letort Spring Run—the double-black-diamond of trout streams—once wrote: "The angler forgets most of the fish he catches, but he does not forget the streams and lakes in which they are caught."

Charlie is right. Recently, I found a photo of a trout I released. It was not an old picture, taken just a couple of years ago on the East Branch of the Delaware. It shows a fine brown—maybe eighteen inches long—lying on its side in shallow water next to my boot for scale. The thing is, I can't remember a thing about it other than the river I caught it in. The trout has melded into the pot of other various nice fish caught over the years but without further incident.

Then there's the tiny scar above my right eyebrow. Every morning when I look in the mirror, I see it. It's from a blackfly bite that got infected during a trip to Alaska. It started to swell on the penultimate day of fishing. By the time I flew home, it was the size of a golf ball and everyone wanted to know if I had gotten in a bar fight with an Alaskan lumberjack.

Today, when I look at the scar, I see a glacial-colored river where the blackfly probably hatched. I can smell air sweetened with the tang of Sitka spruce. And I see the school of silver salmon holding in the logjam. They gang-rushed the purple Egg-Sucking Leech every time I twitched it in front of their noses. Looking back, it makes the subsequent doctor's visit and antibiotics worth it. And it definitely beats another boring story of some schmuck who loaded up on silvers in Alaska.

I sometimes wonder if we anglers didn't occasionally fall in, or get chased off the stream by the bogeyman, or wind up with an ice pack on some bleeding wound, we might just as well take up golf or tennis or some other sanitized sport. Maybe it's the adventure, and risk, or the agony of defeat that keeps us coming back for more.

There's an old joke about a salmon angler who dies and finds himself on the most beautiful river he has ever seen. In his hand is the finest salmon outfit ever made. Standing next to him is a ghillie who instructs him to cast into the most perfect pool imaginable. So he does and immediately hooks a forty-pound Atlantic salmon that gives him the greatest fight he has ever experienced. He lands the fish and the ghillie instructs him to cast again. He gladly does. Now he hooks a fifty-pounder that gives him an even more spectacular battle.

When he releases the second fish, the ghillie tells him to cast yet again. The angler is starting to feel a bit fatigued but how could he resist? Now he hooks a sixty-pounder that gives him the most incredible fight yet. By the time he lands his third salmon he is content and exhausted.

The ghillie tells him to cast another time. The angler declines saying he's had enough for now, that he would prefer to just rest and enjoy the moment.

But the ghillie insists.

Now the salmon angler becomes annoyed and says he really doesn't want to fish anymore, and that the ghillie should leave him alone. Then he adds that since he's in heaven, he should be able to do what he wants.

The ghillie cracks a devilish smile and says, "Heaven!? Who said this is heaven? NOW CAST AGAIN!"

So to my fellow anglers, I say this: Keep making those spastic casts into trees. Go ahead and hook your buddy's armpit, or trip and snap another six-hundred-dollar fly rod. Fall in a few more times. And if you catch something along the way, that's OK, too. Just stay clear from the hell of endless perfect days filled with boring stories and large, forgettable fish.

A quick note about the illustrations: In 2012 I was given a watercolor set as a gift and began painting fish I had caught. Some were caught, photographed, released, and painted; others were caught, photographed, painted, and eaten (not necessarily in that order). In any case, I had fun with it.

Part I

EARLY, SPASTIC CASTS

Jefferson Lake Bluegill

Lot G

The neighborhood bait and tackle shop has largely gone the way of the village soda fountain or corner five and dime. Yet not too long ago, it seemed most towns had one good bait shop or at least a sporting goods store that also sold some hunting and fishing gear. Then came the big box chain store, giant mail-order retailers, and the Internet, and today you have as good a chance of finding the town tackle shop as you would a haberdashery.

In the early 1980s when I was in high school, I used to frequent a local tackle store in my hometown in suburban New Jersey. It was small, but surprisingly well stocked, selling everything from dry flies to billfish lures. But it also had the feel of a locals-only tavern or a men's club, of which I was definitely not a member. A group of regulars—all friends of the owner—always seemed to be in there leaning against this one glass counter full of expensive reels. They were in their thirties and forties and drank coffee or sometimes sipped cans of beer in paper bags. These were the "sharpies"—the so-called 10 percent of anglers who catch 90 percent of the fish. On the wall behind the cash register hung pictures of them holding fifteen-pound steelhead, fifty-pound striped bass, and five-hundred-pound giant tuna.

And to me, the sharpies were gods.

I longed to be accepted into their inner circle. But whenever I walked in the store, they would stop their raucous fishing stories mid-sentence and begin mumbling in hushed tones.

The owner—also a sharpie—sometimes took pity on me and would occasionally give me the briefest of fishing tips. These would come in the form of a knowing nod when I bought a certain lure, as if I was getting closer to finding some treasure that he and his fellow sharpies had hidden from me.

One day I walked in the store to buy still more tackle—maybe a few more Mister Twister jigs—you can never have too many. Or that new Jitterbug painted like a red-winged blackbird, the one that the "biggest kind of bass can't resist," according to an article I just read. I sheepishly slunk past the regulars who were silent as usual and began poring over this incredibly important next purchase.

Then the owner said: "They're catching weakfish at Lot G."

I turned around and noticed he was talking to me.

"What?" I asked.

"They're catching weakfish at Lot G on Sandy Hook. Nice ones," he said again.

The regulars stood behind him in silence watching me.

My mind raced with excitement. I knew where Sandy Hook was—it was a legendary haunt among surf fishermen, an oceanfront peninsula about an hour away known to produce giant fish. I had fished there before and caught a small few bluefish and fluke, but never anything as exotic as a weakfish, another glamour species targeted by the sharpies. Pictures of them holding sleek ten-pounders hung next to the tuna and steelhead. They called them "tiderunners"; they were beautiful and trout-like with purple spots and broad tails.

Trying to sound casual, I asked: "What are they catching them on?" as if I already knew the answer but just wanted to double check.

"Usual stuff. Bucktails. Swimmers," the owner said.

I bought handfuls of each.

"Let me know how you do," was the last thing he said.

Later that day I headed for Sandy Hook loaded for bear with every piece of surf gear I could fit in my car. I didn't know where Lot G was, but by God, I was about to find it.

Not only was I excited about the prospect of catching a weakfish, more importantly, I may have just been let into the sharpies' inner circle. Pretty soon they would be asking me to go giant tuna fishing. It must have been the red-winged blackbird Jitterbug that did it.

When I got to Sandy Hook, I stopped at a ranger station and got directions for Lot G—which turned out to be a beachfront parking area halfway up the peninsula.

I found the lot; dozens of cars were already parked there. Yet the vehicles weren't the usual assortment of beat-up pickups and four-wheel drives most of the regular surf fishermen drove. Instead they looked like typical cars you might see on a summer day at the beach. But this was

October. And although it was a beautiful Indian summer afternoon, it seemed too late in the year for sunbathers. Something began to feel odd.

The weakfish beckoned; time to rig up. I came prepared for a long day—and possible night— of fishing the high surf. I put on a pair of heavy chest waders and foul-weather top. I cinched it tight with a thick canvas wading belt in case a big wave tried to knock me over while fighting a tiderunner weakfish. A sheathed fillet knife hung from the belt to clean my catch. I loaded my surf bag with the new bucktails and swimmers I had just bought, then slung it over my shoulder. I wore a baseball cap and polarized sunglasses with a miner's lamp hung around my neck for later.

Then I grabbed my ten-foot surf rod and headed down a dune trail leading to the ocean. Once again I was surprised—this time by the number of people walking along the trail to and from the beach. People carried beach chairs and wore flip-flops like it was the Fourth of July. Many of them smiled as I walked past. This was getting weirder.

The trail ended at a wide beach. Most of the beachgoers headed to the right, and indeed there were well over one hundred people maybe two hundred yards away in beach chairs and walking along the sand. I barely glanced at them and headed in the other direction to a point where a wooden bulkhead jutted into the ocean. A gentle surf rolled over it and washed into a deep hole. Though I knew nothing about how to catch a weakfish, this seemed like as good a spot as any to try my luck.

I waded into the waves and chose one of the new bucktails. I clipped it to my shock leader and began casting and retrieving it through the deep hole.

But something still didn't feel right. It wasn't the fishing; it was the beachgoers. There was something strange about them. I made another cast then turned while I reeled and watched them one more time. Then I saw something that made me squint like I was seeing a mirage. No, it couldn't be. I shook my head and squinted some more.

Yes, it was. They were all nude.

I was casting for tiderunner weakfish in the middle of a nude beach. I had once heard that Sandy Hook had one, but dismissed it as urban legend. It was not. Nude people lay on towels. Nude people sat on beach chairs. Nude people frolicked in the surf. There was a nude volleyball game going on. But this was not some fantasy nude scene of giggling co-eds conjured up by a hormone-induced eighteen-year-old kid. This was nudity of the masses with every shape, size, and body type on full display. Most of them looked like the kind of people you would see standing on line at the supermarket or the post office—except of course, these people were naked.

And as I gawked at the spectacle unveiling in front of me, some of them began to approach to watch me fish.

I turned back quickly and made another cast, trying to act casual, as if I fish next to naked people all the time. But I found it extremely difficult to concentrate, plus I suddenly felt overdressed by at least four layers beginning with my underwear. After just a few more casts, I reeled up my line, and without ever turning away from the unpeopled ocean, managed to sidestep past the naked beachgoers and back to my car.

Needless to say I didn't catch a weakfish.

As I headed home, I contemplated what had just happened. I wasn't sure whether I was the victim of a practical joke, or had the owner given me a genuine tip but left out some essential fact such as the weakfish were only hitting after midnight? All I knew for sure was that the next time I visited the tackle shop I would be asked for a fishing report.

I waited as long as I could bear it, but a week later I found myself back in the store buying more lures. This time it was a smallmouth jig so deadly it was banned in six states. I was more awkward than usual, hoping the owner had forgotten about the tip he gave me. He rang me up and handed me my change. Then, as I turned to leave, he said, "Hey, how was the fishing down there?" The regulars stood in their usual spot sipping coffee and watching me.

I turned back, gave them all a knowing nod, and walked out.

The Great Lost Fish

Why, oh why, do we pine over lost fish? Why do fish that we only get a mere glimpse of—a coppery flash under a boat, a heavy swirl in a river—remain seared in our memories often far longer than something we actually catch?

The first fish I ever hooked, I lost. It all began when my older brother and some of his friends returned from a duck pond up the street with a small fish swimming round and round in a bucket.

"These are really hard to catch," he said with a fourth grader's earnestness. "They're called sunnies."

He was right, especially since he had caught it on a huge bluefish hook rigged to a heavy boat rod he had found in my parents' garage. I still don't understand the physics of how that poor sunfish managed to wrap its pea-sized mouth around the bait and get hooked.

All I knew is that now I wanted to catch one—badly. So I convinced my mother to take us to the local sporting goods store the next day to get some real tackle. This wound up consisting of a spool of fishing line, a few plastic bobbers, and a dozen hooks. The hooks came in a package with a guy in a red and black checked shirt holding a stringer of trout. How could we miss?

Later that day, I walked to the pond armed with a new, homemade fishing rod: a broomstick with a few yards of line tied to the end. For bait, I had a slice of Wonder Bread wrapped in tin foil jammed into my pocket.

I picked my way around the shoreline periodically lowering the bobber with a tiny ball of bread threaded on the hook. Nothing happened. Eventually I came to a spot next to some lily pads that looked particularly fishy. I lowered the rig and remained still. After a few seconds, the bobber stood on end like an explanation point then plunged under the surface. I lifted, and a five-inch fish just like my big brother caught came shimmering out of the water.

But I didn't know what else to do except stand there with my catch still dangling over the pond. So I jumped up and down yelling to the world: "I caught a sunny! I caught a sunny." Plop. I lost a sunny. That was 1972.

Since then, I've lost hundreds of fish—maybe thousands. Some were just inches out of my grasp like that whipped twenty-inch wild brown hooked on a small New York stream. I was about to grab the leader when it turned its head, spat out the fly, and swam away. With others I never even had a chance—the twenty-pound striper that blasted a popper off the tip of a jetty on a borrowed spinning rod better suited for largemouth bass. I stood barefoot in shorts and a T-shirt and watched helplessly as the bass powered into the rocks and broke off.

Others weren't even my own fish. My brother once hooked a large summer flounder off a pier at the Jersey Shore. When he tried to haul it onto the dock, the line snapped. I watched— horrified—as the flounder sank slowly into the depths. I can still see it fading away like a dream. I was twelve.

Some lost fish made me throw my rod like an angry golfer missing a game-winning putt. With others I just shook my head in a strange combination of disgust and wonder.

Out of all of these, here are the ones that stung the most.

The Bass

At fifteen years old, I was a walking *Field & Stream* article gushing to anyone who would listen about "gin clear water" and schools of fish "stacked like cordwood."

All I wanted to do was catch a so-called "trophy gamefish"—one that leapt and tail-walked. I imagined how I would bring it to a tackle shop to weigh and get a Polaroid taken that would get tacked to the "bragging board." But for all my bluster, my largest catch to that point was a ten-inch yellow perch.

My non-fishing parents, not sure what to do, drove me to a reservoir one spring morning. They sat in the car and read the Sunday paper while I walked the shoreline casting my favorite plugs, spoons, and spinners. After an hour, I had caught nothing.

Then I decided to try a new lure a friend had recommended: a purple plastic worm with a strange hook-shaped tail. He warned me not to try to thread the hook through the tail as he had first attempted, but instead I was instructed to thread a jig-head through the other end. The tail, he explained, would corkscrew through the water like a swimming minnow. I tied on the lure, made a short practice cast, then watched with delight as the worm came to life snaking and swimming along the lake bottom. In *Field & Stream* lingo: "it had that come-hither action that no gamefish could resist. . . ."

Now I began working the shoreline with renewed confidence. Even if I didn't catch anything, it was fun just watching the worm with its amazingly realistic swimming action. I would cast and let the lure settle watching coils of line slowly peel off the spinning reel. Eventually, the line would stop—indicating the worm had reached bottom.

But on one cast, it seemed to take longer. Then I noticed the line wasn't actually sinking but moving parallel to the shoreline. Confused, I reeled until the rod bent deeply.

Now comes the image I will never forget: The line slowly rises as the rod surges toward an unseen force. The water humps up, then parts. A massive black hole of a mouth followed by the dark green bulk of a humongous largemouth bass launches from the reservoir and yes tail-walks across it. I had hooked my trophy.

The fish jumped a second time and a third. I had no idea what to do. The bass was out of control and now I wanted it to stop. I felt like the best-looking girl in high school had just pulled me onto the dance floor at the prom to do the merengue. And I can't dance.

It ended as suddenly as it began with the bass throwing the plastic worm on its next jump. I stood still for a long time. Eventually I walked back to the car knowing that no matter how much I explained it to my parents, they would never understand the magnitude of what had just happened.

To this day, I have no idea how big that bass was: two pounds; ten pounds. It doesn't matter.

The Shad

Hooking any shad feels improbable. Why would a large sea-going herring, sex-crazed and not feeding, have the temerity to chase down and strike a shiny doo-dad in a freshwater river many miles from the ocean? But they do, and that's why I cast for them.

The season before, I had landed my first-ever: a small buck that fought hard but weighed only a pound or two. It reminded me of the cocktail bluefish I would sometimes catch in the surf—minus the sharp teeth. The larger females—egg-laden roes: thick shouldered, deep bodied, and two or three times heavier—remained elusive.

That is, until one late April day on the lower Delaware River. I had waded to a rocky perch to reach some heavy water above a set of rapids. There I cast a red-and-white shad dart into a deep, surging run where I had guessed some fish might be resting.

The river roared wildly around me. If I slipped here, they might find me a week later in a logjam. As I fished the run, I quickly learned that the bottom of this part of the Delaware might be magnetic as it seemed to pull virtually any lure I cast into a crag where it got stuck. In thirty minutes, I had already paid with half a dozen lost darts.

So when the rod bent again and I lifted into something immovable, I had assumed it was simply another hungry Delaware River rock.

Except this rock moved.

The rod lurched down-current. Drag slipped off the reel, slowly at first, then began accelerating. Even above the din of the rapids I could hear the ratcheting sound rising in pitch to a final zzzziiip!

Then, framed by the white haystacks of rapids, the chartreuse foliage of early spring, and a blue April sky, a heavy silver shad catapulted three feet above the river. The image looked so perfect, so beautiful, it seemed like one of those corny, stylized calendars of a trout arcing out of the water and a grinning angler in hip-boots and a fedora holding a bamboo fly rod.

But this was real and I was grimacing, not grinning. The fish crashed back in, and my mind raced trying to determine the physics of how to land it. A five-knot current, super-slick boulders, and six pounds of freaked out anadromous fish stood between me and glory.

The shad made it easy on me. It streaked downstream into the rapids parting the line as an afterthought.

I have since caught many shad. Some fought as hard; some jumped as high; some may have been larger. But that one leaping roe still hangs above the Delaware—a silvery apparition forever out of reach.

The Striper

This incident begins with the only superhero I ever saw. I was casting for striped bass at daybreak along a row of rock jetties during a nor'easter. Despite a dozen previous trips, I had yet to even see my first surf-caught striper. With every skunking, they were taking on near-mythical status.

Now with a stiff headwind making my plug die mid-cast, I knew I was in for more of the same.

The jetty in front of me looked as inviting as it was treacherous. Breakers crashed into the windward side sending plumes of whitewater into the air. A strong current rushed along its length creating a sort of flowing river that ran out to sea. Migrating baitfish blocked by the jetty would get swept to where stripers surely awaited. But fishing from the rocks was impossible. They were covered with slimy black algae that may be the most slippery substance on the planet. "Snot on glass" is how I once heard it described. The rubber lug soles on my chest waders would be like a pair of roller skates if I dare set foot on a single rock.

"Morning!" I heard someone say behind me. I turned and saw him. He wore rubber coveralls and a foul-weather top cinched with an army surplus belt. His face remained mostly hidden by a hood. On his boots he wore heavy rubber cleats with steel carbide studs. He held a custom surf rod rigged with a well-used wooden swimmer. A canvas plug bag bleached pale from saltwater hung from one shoulder, and across the other hung a large gaff fixed to a long aluminum handle.

He walked past me and stepped onto the jetty. He clicked along the slick rocks graceful as a snow leopard strutting up the Himalayas. Halfway out, he fired a cast into the churning surf. He reeled slowly and carefully, ignoring the waves splashing around him. He fired again, reeled a few turns, and then reared back suddenly on his rod, which bent and throbbed. Minutes later he grabbed his gaff, and with a mighty swing brought a fifteen-pound striped bass onto the rocks. He didn't stay much longer after that. He held the bass by the gill plate, walked back along the jetty, then hopped onto the sand. He sauntered past me giving me a nod. I nodded back, speechless. Maybe it wasn't a superhero but God himself.

Later that day, I made the owner of a local tackle shop very happy by purchasing jetty cleats, foul-weather gear, and a plug bag. I borrowed a gaff from a friend's dad who owned a boat.

That same night, I ventured onto the jetty with two friends. The wind had dropped out leaving just a mild swell that gently rolled and gurgled along the rocks. My new cleats dug through the algae gripping like mountaineering crampons. My friends were not so lucky, slipping and crawling over slick spots until all three of us finally came to a higher, dry outcrop near the end.

We spent the next two hours casting and casting. We switched lures and varied our retrieves but caught nothing. We were beginners; the idea of moving down the beach to the next jetty never occurred to us. Someone had caught a striper at this very spot on this very morning, so by God, we were staying.

To pass the time one of us would occasionally yell out: "I got one!" causing the other two to jump and scramble about. Then they would realize a joke had been played and curse at the prankster who would laugh idiotically.

It was well past midnight when my plug stopped dead just beyond one of the far rocks. I lifted the rod and could make out churning white-water in an otherwise calm sea. I had hooked a striped bass.

"I GOT ONE!" I cried out triumphantly. "GET THE GAFF!"

Both of my friends laughed. They weren't falling for the same trick again. I had become the boy who cried bass.

I turned to one of them and pleaded: "I REALLY HAVE ONE! GET THE GAFF!" I gestured wildly and took my hand off the reel for the briefest of moments. That was all it took to give the bass a few inches of slack so it could toss the plug. And it did.

They continued to laugh like idiots still not yet believing I had hooked anything. I gnashed my teeth and stomped my cleats leaving angry gashes in the jetty rocks—a petroglyph of frustration that I'm sure is still there.

The Trout

I bought the fly rod off Herman—an Orvis Superfine, the kind with unsanded little ridges running down its length and a soft sultry action. I had spent the summer learning to cast it in the backyard; now it was time to take it fishing. The plan was to canoe and camp along the Upper Delaware stopping when I felt like it to fish. But, if the fly rod didn't work, I would quickly ditch it for a spinning rod and a Rapala or a Phoebe. I was still in the stocked trout phase of my angling career; I remained skeptical of fly fishing as a silly, ineffective way to fish.

Still I wanted to give it a try. So upon the recommendation of a friend, I stopped at Dette's Fly Shop in Roscoe. Dette's has been there since 1928, and unbeknownst to me at the time, is hallowed ground for fly anglers. Mary Dette, the daughter of legendary fly tiers Walt and Winnie, carefully filled a plastic container for me with beautifully tied dries and wets. She named each pattern as she placed them in the cup:

"Delaware Adams, Cahill, Elk-Hair Caddis, Yellow Sally." I nodded dumbly eyeballing a few dusty Phoebes for sale hanging on the wall.

Later that day, after I had pitched my tent and set up camp, I sipped a beer and gazed at the river considering my fishing options. Maybe I would cast a plastic popper for smallmouth or dig up some worms. Just then, I heard a small splash downstream as if an acorn had fallen in the water. Then I heard another. When I saw the third, I realized the splashes came from fish. I looked closer and could now see small insects trickling off the river. I realized I was not only looking at my first mayfly hatch but also wild trout rising to eat them.

I chugged my beer then tugged on my waders. I looked at the spinning rod briefly, but grabbed the fly rod instead and stuffed the cup of new flies into my shirt pocket. Then I waded in and slowly approached the rising fish. The trout held in a narrow slot of deeper water running hard against the bank. Once or twice a minute, one of them would come up and take an insect in a deep gulp.

I chose my largest fly, going by the bass axiom of big bait/big fish. But several casts later, it remained untouched. It quickly became apparent that these fish were different. Then a small insect flew past me. I scooped it up with my baseball hat and examined it. It was yellow. Then I opened the cup of flies and found one that looked similar—the Yellow Sally. More pieces of the puzzle were falling into place.

I tied on the little yellow fly and began false-casting it. I aimed for a particular fish that was rising steadily under an overhanging tree limb.

Whenever I think back to what happened next, I like to imagine the theme of *2001: A Space Odyssey* playing in the background. I make the final cast and shoot the line in slow motion as the first two rising notes play: Daaaaa . . . Daaaaa . . . the leader unrolls and the fly touches down on the third note—Daaaaaaaaaa. . . . The trout comes up, eats the fly, and I set the hook confidently. DA-DAAAA!!!!

I had made the quantum leap. I was a dry fly fisherman.

But instead of the triumphant crescendo of *2001*, now *Flight of the Bumblebee* plays. For the sixteen-inch rainbow trout on the other end of the line goes absolutely bat shit. It leaps three times in as many seconds, then runs upstream, downstream, upstream again, then downstream one last time before snapping the leader. Maybe five seconds passed from when I first hooked it. I reel up slack fly line with shaking hands. I gaze at the fly rod and for the first time realize it has an "on" switch that temporarily turns it into a light saber.

From that point on, I knew two things: first—I would chase that rainbow trout forever; second I would never again call fly fishing silly.

All these years later, these four fish still haunt me. Sometimes I think of what I could have done differently to land them: a better knot, more drag, less slack. Would things be different if I landed them?

Recently, I was talking to my brother about our vacations at the Jersey shore when we were kids. I brought up the giant flounder he lost off the dock many years ago. He looked at me oddly.

"What flounder?" he asked. He was never much of a fisherman, so I retold him the story in excruciating detail—the line snapping, the fish fading slowly into the depths. But the more I explained, the more I realized he had completely forgotten it had ever happened.

In a way, I envy him.

The Diamond Jig

It was my dad's idea: "Go buy yourself a lure so you won't need to fuss with all that bait." This advice came after watching his teenage son spend the better part of a family vacation at the Jersey Shore setting traps for killifish, which in turn I would use to catch snappers and fluke.

My father was not a fisherman nor an outdoorsman, so he did not know the joys of pulling a killie trap loaded with shimmering baitfish. Nor did he know anything about lures. To him, a New Yorker raised on urban shores better known as the Upper East Side of Manhattan, it seemed like a pragmatic way to avoid handling any sort of bait, alive or dead.

At the time, I knew nothing about lures either. My fledgling tackle collection consisted of a few hooks and sinkers, period.

Nevertheless it did seem like a good idea. After all, keeping bait-fish alive was difficult without expensive aerators; and transporting any bait—alive or dead—was messy business. Earlier that summer, I learned this firsthand when I left a fluke rig baited with a piece of squid in my tackle box. A week later, I opened the lid, and an explosion of stench punched me in the nose. No matter how much I scrubbed and disinfected it, years later that tackle box still smelled faintly of rancid squid.

Then one afternoon in August, before I could drive, my parents agreed to take me to a fishing pier on the northern New Jersey coast. They even said we could stop at a tackle shop along the way. And there I would buy my first-ever lure.

We drove to a salty place known to cater to serious fishermen. The staff were notoriously ornery; they expected you to know exactly what you wanted when you walked in the door. Newcomers and beginners were not particularly welcome. Tackle Nazis.

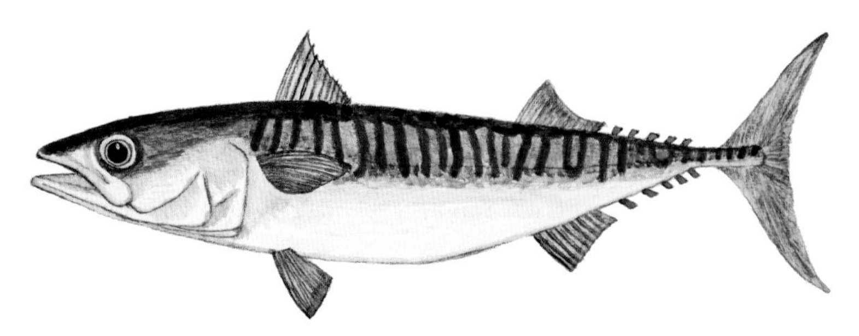

Atlantic Mackerel

My parents handed me five dollars, then waited in the car as I walked in alone. The shop smelled like a combination of cigarette smoke and eels. Two women sat in chairs at the back of the shop—sisters of the owner I later learned. They were counting and boxing live sandworms by the dozen. Lucky Strikes dangled from their mouths. Above them, gigantic, stuffed striped bass hung from the walls covered in a yellow patina of several decades of tobacco smoke.

I sheepishly asked where the lures were, and one of the sisters pointed to a far aisle. I turned the corner and stood transfixed. Dozens, no hundreds, of lures of every shape and size hung from a massive wall of pegboard. Bunker spoons larger than any fish I had ever caught; wooden plugs as long as my forearm; brightly colored surgical tubes with enormous hooks dangling from them. Poppers, chuggers, tins, swimmers, bucktails, deep divers, shallow runners. The brand names sounded like weaponry: Polaris, Atom, Bomber. How could I possibly know which one to use? Asking the Sandworm Sisters seemed out of the question.

Then I saw it: a lure that promised glory. The package screamed out in bold letters: "STRIPERS!" "BLUEFISH!" "WEAKFISH!"—noble gamefish I had never caught before. Inside the package was a slender chrome-plated beauty about six inches long with a single hook dressed with a smart yellow feather. I was looking at my first diamond jig. Forget the gamefish; this fisherman was gut-hooked. I happily handed my money to one of the sisters who took it with fingers stained like the striper on the wall.

We arrived at the fishing pier—a relic from another era that was once a steamship port. Most of it had since been converted to an amusement park. Just a few years later, the entire structure would burn to its pilings and eventually be demolished.

To get to the fishing section at the end, you walked through into a combined tackle shop, luncheonette, and bar then paid an admission fee. Regulars sat at the counter eating sandwiches or nursing their first beer of the day. The operation was run by a very large, tough woman whom the regulars called Dolly. She did not look like a Dolly.

After my parents paid my admission, Dolly asked me if I needed any bait. "No thanks," I proudly announced. "I've got a lure."

I ventured to the end of the pier. Forty feet down, a gentle swell surged and hissed at concrete pilings encrusted with mussels and barnacles. Dark purple colors below the surface marked a submerged reef that dropped into mysterious blue water.

A few fishermen were scattered along the railings. No one seemed to be catching anything. My parents sat on a bench and started reading the newspaper.

The moment had come. I tore open the package and took out the diamond jig, which, true to its name, gleamed and shined in the summer sun. I tied it on, its yellow feather fluttering in the warm breeze.

Then I cast out as far as I could. The jig sailed over the ocean like a chrome missile then seemed to hang in the air for a second before plunging into the Atlantic with a muffled splash. Line poured off the reel as the lure dove deeper and deeper. Eventually, I felt a dull thud that meant it had reached bottom.

Before I describe what occurred next, it's important to know that all anglers trace their beginnings back to the primordial ooze of not knowing what the hell they are doing. They do things like crank spinning reels upside down with the wrong hand, try to squeeze into a pair of chest waders without first taking off their shoes, and scratch their heads wondering how three fish can fit on those weird three-pronged hooks all at the same time. Some of us slowly climb ashore and evolve; others languish for their entire—and probably brief—angling careers.

Now with this disclaimer out of the way, I shall explain what happened: After I made that first cast and let the lure settle to the bottom, I leaned the rod against the rail, sat with my parents and waited for a bite—just like I would with a big piece of bait. You see, no one had told me, and nowhere on the lure's package did it say you had to actually move it to make it work. I thought a fish would simply swim up to the shiny jig and eat it as it lay motionless on the bottom.

This most basic truism of virtually all lures—that they must be jigged, popped, chugged, twitched, crawled, or otherwise retrieved—remained unknown to me.

All three of us sat on the bench. Periodically, my parents would look up from the newspaper and ask me if I was getting any bites. I wasn't.

After a while, I reeled in the diamond jig and cast to a new spot. Maybe there was a big striper or bluefish lurking over there that would find the lure. Nope. Maybe a school of weakfish were over there. Not a chance. Mercifully after about half an hour, I managed to snag the jig in the reef and couldn't pull it free. I had to break the line, leaving my first-ever lure at the bottom of the Atlantic. When that happened, my parents gave me another few dollars, and quickly I retreated to the tackle shop, bought some frozen spearing, and caught some snappers.

But it turns out my dad was right about lures. Eventually I emerged from the ooze and learned how they worked through trial and error. I now have tackle boxes packed with them: spinners, spoons, plugs, and of course, diamond jigs, which, as the package promised long ago, do catch fish. But only if you know this secret technique: You cast it out and then you reel it in.

The Chain Letter

You don't see chain letters much anymore. They would come in the mail in an envelope with unfamiliar handwriting and no return address. Inside would be a blurred copy-of-a-copy-of-a-copy explaining that you are now part of a magic chain that mustn't be broken.

The letter promised good luck in exactly one week, but only if you made ten copies and mailed them to ten friends. More importantly, it warned of bad luck if you broke the chain. The letter would list well-documented examples of both scenarios: "Robert S. did not break the chain and won the lottery." And: "Doris G. ignored the chain and died in a horrible car accident." And then there would be a redemptive example: "John R. broke the chain and lost his job; but then he later completed the chain and got a better job."

I received a chain letter when I was eighteen or nineteen years old. I was still living at home and was probably leafing through the mail looking for my latest *Field & Stream* when I found a strange envelope addressed to me. My friend Adam happened to be over at the time.

"What the hell is this?" I said, and then read the letter out loud, snickering between sentences.

When I was done, I started to crumple it up when Adam grabbed it from my hand.

"What are you doing?" he asked in disbelief.

"Throwing this crap in the garbage," I said.

"Are you insane? Didn't you read this?" He read the letter again raising his voice to emphasize certain parts: "DORIS IS DEAD. HELLO? You can't break this or you're screwed."

We must not have had a lot going on that day, because the next thing I knew, we found ourselves at the local stationery store buying ten envelopes and having ten copies made. Then we went to the post office to buy ten stamps. We picked our "friends" by randomly going through the phone book. Finally, the letters were mailed. The chain remained unbroken—at least from me.

"You did the right thing," Adam said reassuringly.

I felt like I got cheated out of three and a half bucks.

A few days after I mailed the chain letter, I went fishing at a local pond up the road from my parents' house. The pond was "new"—part of an impounded swamp and adjacent woodland made into a nature park the year before. There were rumors it had since been stocked with largemouth bass and black crappie.

But so far, at least for me, the pond remained an enigma. It looked fishy with a brush-covered shoreline and dead trees sticking up here and there. But the few times I had fished it with lures, all I caught were a few small sunfish.

This time however, I had set a minnow trap overnight. When I pulled it, a few baby bluegills flipped around inside. I threaded a silver-dollar-sized fish on a hook rigged to my ultralight spinning rod, then flipped the bait just beyond a dead tree sticking up six feet from shore.

The bluegill descended a few inches then suddenly vanished in a large silver flash. Line began rapidly pouring off the reel. I wasn't sure what to do—let the unseen fish keep taking the bait, or set the hook before it realized something was wrong.

I couldn't take it anymore. I flipped the bail with a click then heaved the rod back into a deep and satisfying bend. There was a boil and churning on the surface and another brilliant flash of silver. A gigantic crappie wallowed and rolled and eventually allowed me to lip it.

Measured against the ruler built into my tackle box, it taped a full fifteen inches and was as broad around as a pie tin. Covered in black speckles as big as dimes with a mouth that could swallow a golf ball, it was twice the size of the biggest crappie I had ever seen. I released it, knowing I had cracked the code on this pond, that it did in fact contain some very large fish.

For a few sweet years after that day, the pond would become one of my most prized secret fishing spots.

And yes, this happened exactly one week after I heeded Doris's cautionary tale and did not break the chain letter. I only wish I could have somehow contacted its original author and proudly offered my own testimony to the power of its obvious magic.

Getting Thrown In

Let's say you are a scrawny nineteen-year old, and you have a friend named Herman who is the same age but six feet five inches and 265 pounds of volatile German/Haitian stock. And let's say Herman has never had a girlfriend and is absolutely raging with hormones; and due to these two circumstances, he has decided to substitute pleasures of the flesh with pleasures of the fin. Here's some valuable advice: Don't get in the way of his fishing.

Especially if he has backfilled huge gaps in his life with a dump-truck load of fishing gear: dozens of rods and reels ranging from delicate fly outfits to heavy saltwater rigs; tackle boxes of all sizes filled with every known plug, jig, spinner, spoon, and fly; waders, boots, and foul-weather gear suitable for a commercial swordfish boat; nets and gaffs that could capture bluegills to small tuna; and a single electric trolling motor even though he doesn't own a boat.

This warning holds doubly true if it's a beautiful Friday afternoon in late May and Herman has just been sprung from his awful job as a mason's helper and the only thing that has got him through the week was the thought of fishing for trout on a stocked New Jersey stream with his new Orvis fly rod. On top of that he has just paid another mercenary "friend" an outrageous fifty dollars in gas money to take him there since he doesn't own a car.

Just let him fish, even if you want to head home because it's getting late and you haven't caught a thing with your spinning rod, and you're starting to get bored because you have the patience of a nineteen-year-old, and you start to resent both your mercenary friend and Herman for a cash transaction that has led you to this mediocre New Jersey trout stream.

And do not complain incessantly as Herman continues to cast with the resolve of a great blue heron even though he is fishless, too, while the other friend just sits smugly with fifty bucks in his pocket.

And finally, whatever you do, never—ever—rig up a buzz-bait the size of a small outboard motor and cast it noisily and spitefully over the deep trout pool Herman is trying to delicately fish with his new Orvis fly rod. And if he warns you at least three times to stop, for the love of God do what the man says.

Because if you ignore these warnings, you just might find yourself in a bear hug that makes you feel briefly like Fay Wray in King Kong's grasp just before you get tossed headlong and flailing into said trout pool. The water will feel particularly cold.

And when you get out of the pool seeing only red, don't grab Herman's massive tackle/fly box and start to throw it in the river, because he will simply pick up your own tackle and threaten to do the same. This will force you to stand down. All the while, the mercenary will snicker at the whole situation and with fifty dollars still in his pocket.

And then you'll stand there humiliated, dripping wet while Herman continues to cast for as long as he likes, even though he knows every trout has long since been spooked when the skinny, obnoxious kid went flying into the pool.

You have been warned.

Ya Gotta Hook 'em to Cook 'em

My best friend Scott was not a fisherman. He went with me a few times just to watch, and even saw me catch my first-ever striped bass from the surf when I was twenty. I landed the fish—about an eight-pounder—and proceeded to jump around the beach like I won the lottery. He stood there embarrassed for me, probably relieved there were no girls around.

Scott's dad was from Long Beach, in Queens, New York. He wore lots of gold jewelry and drove a Cadillac. One time, he took his family out to dinner and invited me to come along. Somewhere during the meal he asked me about fishing.

The conversation went like this:

"So Scott says you're a big fisherman."

"Um . . . yeah."

He lit a cigarette and said, "When Scott was little he kept asking me to take him fishing. So I bought a pole and some tackle, and we fished off the causeway in Long Beach."

Scott nodded, remembering the day with a not-so-fond expression on his face. "Yeah," Scott said, "We sat there all day and never caught anything. All these other guys were catching fish. We used those weird worms that bite."

I knew immediately what he was talking about: bloodworms. And they do bite. But they are also excellent bait. I could picture the scene: They were probably fishing for porgies or winter flounder. Guys lining the bridge bailing them two at a time. Fish flying over the rails and flopping on the bridge deck. Blood and scales splattered around. Buckets full of glistening fish. In other words, heaven.

Scott's dad took a long pull from his cigarette, his gold bracelet glinting in the restaurant light. Then he turned to Scott and said, "I never baited your hook."

"WHAT!?" said Scott.

"Never baited your hook." Little puffs of smoke followed each word. "I wasn't going to touch those disgusting worms."

There was a pause as if a potentially tasteless joke had been told. But then the whole family—Scott's mother, his two sisters, and of course his dad—burst into guffaws, followed by Scott who seemed to laugh especially hard.

I chuckled politely, but kept wondering how many fish they might have caught if only his hook was baited. Maybe Scott would have rejoiced on the beach with me—two weirdos instead of just one jumping and celebrating like cavemen around a flopping striped bass.

Fishing Ruined My Life

From the mid-1980s through the early 90s there was an elite breed of fishermen you rarely see anymore. They drove old four-wheel drives—rusting Broncos, Scouts, and Blazers. They had racks of PVC pipe bolted to pressure-treated 2x4s that hung from the bumpers of their trucks. In them, stood twelve-foot surf rods rigged with huge wooden plugs or rigged eels. And they caught fifty-pound striped bass from the surf under the cloak of darkness.

At that time stripers were a mystery fish; their numbers were only beginning to recover from historic lows just a few years before. Few caught any; those who did fished only at night. You would see them sometimes—ghostly, solitary figures casting on a moonless beach or heading back to their trucks at dawn when the rest of us were just gearing up to make our first casts of the day for bluefish or fluke.

When I was in my twenties, I met one of them. It was at Sandy Hook, the northernmost peninsula on the New Jersey coast. We wound up talking for a few minutes while we stood by our trucks.

He was forty-ish and looked awful—swollen, baggy eyes from perpetual lack of sleep, uncombed hair, and a three-day beard. His grungy clothes hung off a physique that looked like it was built from late night doughnuts and lots of coffee.

I happened to have a snapshot of a large striped bass I had recently caught, and I showed it to him. Though I didn't say so, I landed the fish at Sandy Hook's bayside, which at the time was a relative secret, lightly fished compared to the vastly more popular oceanfront beaches. The photo itself revealed nothing. It was taken at night with a flash. The background was pure black, and I was kneeling on a beach that could have been anywhere. I had no intention of telling him where I caught the big fish.

He looked at the photo, grunted, and said: "Yeah, I heard they were catching nice fish on the bayside."

I was astonished: "How did you know it was the bay?"

"You can tell by the shells—you only see shells like that in the bay." I studied the photo. Sure enough, there were tiny clam shells in the foreground that I never noticed. This was a guy who clearly spent a lot of time walking the beach.

He went on to tell me he had been fishing for striped bass all of his life. He said it was all he ever thought about and all he ever wanted to do. He then told me he was divorced and had lost jobs and relationships because of his obsession with stripers.

He was not proud of this. Then he said more to himself than to me: "Fishing ruined my life."

He said goodbye and walked to his truck, an old Bronco that I'm sure smelled of bait and had fish scales lacquered to the seats. Then he drove away looking to catch the next good tide.

I had just met my hero.

Part II
SNAGS

Loon Lake Rock Bass

Opening Daze

The problem with fishing opening day of trout season in New Jersey is that it's opening day of trout season and you are in New Jersey. Yet I have continued to participate in this odd spectacle on and off for better than thirty years.

I rode my bike to my first opening day. My father had cut out an article from the local paper listing all the stocked waters in the state. When I scanned the list, I was delighted to discover a stream in a neighboring town that received a few hundred stocked trout.

The night before the big day, I readied my tackle: a fiberglass fly rod with a black foam grip that I could barely cast along with a box of wet flies I bought at a garage sale for fifty cents. I had never been trout fishing before and honestly expected to see just a few fly fishermen wading the stream and artfully casting—perhaps while thoughtfully smoking their pipes. Maybe they would offer me some streamside casting tips or show me how to properly field dress my catch.

My God, I had no idea what was in store for me.

The next morning, as I excitedly peddled up the busy road that bordered the stream, I began seeing dozens of cars crammed into every possible turn-off. Then I saw the stream itself. Fishermen lined every conceivable pool, pocket, and run. They cast nightcrawlers, salmon eggs, and bizarre fluorescent lures. Some fished strange "grocery baits" like marshmallows and canned corn. They crossed each other's lines and yelled at each other. All used spinning tackle; not a fly rod could be seen.

But they were catching trout, and at that point that's all I cared about.

I locked my bike just as a man with hip-boots dragged his limit of six hatchery rainbows past me. "You're a little late, son," he said before throwing his gear and fish into the trunk of his car and speeding away.

He was right. Opening day in the Garden State starts at 8:00 a.m. exactly. Get there at 8:15 and you've missed the best fishing of the day.

I sheepishly attempted to join the scrum, but wound up relegated to ankle-deep flats. There, I half-heartedly practiced my fly casting before eventually heading home, humiliated.

After getting schooled that first year, I learned how to properly fish subsequent opening days—New Jersey style. The first thing I did was ditch the fly rod, replacing it with three spinning rods (it would be years before I even considered fly fishing again). I fished bait only: worms, canned corn, and a secret weapon called Uncle Josh Trout Cheese. It came in a jar and had the smell, look, and feel of dog crap. But it caught trout and that's all that mattered (not sure what that says about the discerning palate of a hatchery trout). And I would get to the stream hours early to stake my claim and set up my three rods. Anything I landed would get clipped to an aluminum stringer and brought home just like everyone else was doing. I'd fry the bland fish for breakfast and convince myself they were the same trout Ted Trueblood of *Field & Stream* was catching and eating in Montana.

But after a few years, opening day began to lose some luster. Maybe it was catching the same eleven-inch stocked fish every year, or the rowdy crowds, or the funky waterways lined with trash from slob fishermen. Maybe I got tired of washing Uncle Josh Trout Cheese from under my fingernails.

At about this same time, I discovered shad fishing in the Delaware River. Shad, running up from the ocean to spawn, are big and wild and strong—everything a hatchery trout is not. Now in my early twenties, my decision to move on was easy—it was like ditching the plain girl next door for a supermodel. I quickly turned into a shad snob (still am) and looked down on the great unwashed fishers of stocked trout. One of my favorite shad spots happened to be near a popular trout-stocked canal. Sometimes I would slowly saunter past the crowds and their puny trout smugly carrying a five- or six-pound glistening shad I had taken at dawn.

Opening day was clearly in the rear view mirror of my fishing career—or so I thought. Then a few years later, I began taking my nephew Johnny fishing. We caught panfish and bass in some local lakes. But one day I found myself suggesting we ought to try fishing for stocked trout on opening day. I'm not sure why; it felt like a rite of passage he should experience as a fellow angler. So back to the stocked streams I went for a few more seasons. These trips became well-planned assassinations. Johnny and I would fish silently, arriving pre-dawn to grab the best spots. At 8:00 a.m., we would fish ruthlessly, quickly taking our limits. Then we would head home where I would fill my electric smoker. It turns out that five hours and two pan-fulls of applewood later, your standard issue eleven-inch stocker is very tasty.

Last year, Johnny started college so it was time for my nine-year-old son, Finn, and eleven-year-old nephew, Sean, to enter their opening day heritage. Sean had expressed a genuine interest in fishing. Finn, on the other hand, had to be coerced with the promise of unlimited doughnuts and hot chocolate.

I decided to fish a nearby suburban river I hadn't visited in a number of years. It's kid-friendly: slow flowing and heavily stocked. I knew a particular stretch where the three of us could spread out along a short concrete bulkhead.

Again, we readied our tackle the night before, rigging several spinning rods and stacking worm containers in a small cooler. I set the alarm for 5:00 a.m. The goal was to be on the water by 5:30.

When we got to the river, I was surprised to see a lone fisherman standing just upstream. Right in front of us, where we intended to fish, three folding chairs were set up. They took up the entire bulkhead, which was no more than ten feet wide.

I asked the fisherman if the chairs were his, and he said no. Then I asked him how long he had been there. He told me 4:30 and said the chairs were there when he arrived. I looked around; there were no other anglers in sight. I speculated that maybe they were forgotten from the day before or they belonged to someone in one of the nearby houses.

I stood there for a minute not sure what to do. Then the fisherman suggested I should go ahead and set up my gear and not to worry about it. I agreed. I laid the rods on the bulkhead in front of the chairs thus claiming my spot.

Then began the longest wait in all of sports: the arrival of eight o'clock on opening day morning in New Jersey. To pass the time, I chatted with the lone fisherman. Finn and Sean napped on a blanket. Meanwhile, other anglers began setting up nearby and across the river.

Half an hour later, another angler showed up. He was about sixty-five and carried a handful of fishing rods, which he began propping up against each of the lawn chairs. I stood five feet from him watching, but he ignored me. I looked at this new fisherman, the chairs, Finn and Sean, and the bulkhead. There was no way all of us could fit.

That's when I said: "Um, it's gonna be a little crowded here."

He looked up as if just noticing me for the first time. Then he glanced at my tackle and said, "Yeah. Is that your tackle? Better move it. There's three of us fishing here. My friends are right behind me."

The conversation quickly deteriorated into this:

Me: "Sorry. I was here before you."

Him: "No you weren't. I set up these chairs the night before."

Me: "What!? You can't claim a spot the night before with some lawn chairs."

Him: "Yes I can! I do this every year!"

Me: "That's ridiculous. You can't do that. It's not fair!"

Him: "You're an asshole!"

Meanwhile my son and nephew were fascinated by this display of testosterone, which they watched like a hard-fought tennis match. I stopped, took a deep breath, and looked at them both, suddenly horrified at what the boys were witnessing. I foresaw the situation only getting uglier. Clearly this man was not moving. And his friends were showing up any second.

So I announced to the boys we were leaving because of this rude man and we would have to find a new spot. I grabbed our gear, wished "Mister Chairs" a really nice day, and walked away.

By now, most of the good spots were taken. So we wound up crossing a nearby bridge and setting up directly across the river from Mister Chairs, who was now joined by his two buddies who were glaring at me smugly. To make matters worse, the water where we now stood looked shallow and fishless. I was getting flashbacks of my first opening day. I was a little late, son.

But to the boys, they had just seen high adventure and wanted to relive it over and over. So we did. They barraged me with questions: "Why was that man so mean? Why did we leave?" Then they began parroting me: "You can't set up those chairs the night before."

Then Finn, giggling and with jelly doughnut smeared on his face said, "Dad, you should have thrown his chairs in the water."

I admonished him for having such a great idea—I mean bad thought.

Sean, older and more mature, took a scientific approach: "Next year we should put acid on the legs of the chairs the night before so when they sit in them, the legs will break and they will fall in the river."

Finn, now fully jacked up on sugar, cackled: "Next year let's put dog poop on the chairs!"

Sean and Finn howled with laughter.

"Finn! Stop! That's wrong!" I said, now giggling a little myself. Maybe a better idea would be to smear the chairs with Uncle Josh Trout Cheese.

Eight o'clock came and went. The water was cold and few trout were caught. Mister Chairs and his group caught a few. We got nothing. By 9:00 a.m., we headed home.

But this was not a trip whose success was measured by a few measly stocked trout dangling from a chain stringer. For two boys, this opening day would always be known as the year they almost saw a fight with Mister Chairs.

As for next opening day, all I can say is that it's coming soon, and I know just where those three chairs will be.

Party Boat Blues

This story is absolutely, 100 percent true except for one thing: the name of the main character, which has been washed away by the sands of time. But as I have retold the story over the years, his name has become, and will be forevermore, Louie.

Though I was never formally introduced to him, Louie was a fellow passenger on a party boat fishing off the Jersey Shore one July day. I had convinced my then girlfriend to join me for what I billed as a romantic time on the water, which to me meant chumming for bluefish.

Anyone who can pay the fare is welcome on a party boat. Subsequently they can be reminiscent of a crowded DMV but with tackle and bait. They attract tattooed men who drink beer and smoke cigars at 7:30 in the morning; tourists with white socks and sandals who may have never caught a single fish before; guys in rubber coveralls straight from a Gloucester longliner; sunburned women in tank tops who smoke and spit—and guys like me sitting sheepishly next to their slightly pissed off girlfriends. It's an egalitarian mix of beginners and experts, rich and poor, couth and uncouth, toothed and untoothed.

Louie was a party boat regular—a sharpie—the kind of guy who palled around with the mates and often won the pool for the biggest fish. He was burly and tanned, shirtless and bald. And he had a fillet knife in a sheath strapped to his belt like a six-shooter.

Louie had muscled his way to the crowded stern of the boat, often the best spot since that's where most of the chumming takes place. Meanwhile, I set up amidships, knowing the competitive stern is no place for a girlfriend, especially one who, as previously noted, was slightly pissed off. As the party boat steamed its way to the bluefish grounds, I did my best to console her with at least the promise of good fishing.

Laurence Harbor Bluefish

An hour later, it was clear I had broken my promise. Regardless of where you were positioned on the boat that day, fishing turned out to be slow. The mates slung ladlefuls of chum, baits were drifted and retrieved, but few people caught anything.

That is, except for Louie, who seemed to outfish everyone onboard. And everyone knew whenever Louie hooked a bluefish. You would suddenly hear a very loud: "FISH ON! COMING DOWN! GET OUT OF MY WAY!"

As the bluefish fought, Louie would follow it pushing and shoving his way past his fellow fishermen. If they did not reel up their line fast enough, risking a possible tangle, Louie would whip out his fillet knife and with a flick of his wrist and flash of steel, cut their line. Sometimes two or three lines would be cut in the course of Louie battling a single bluefish.

It quickly became clear that Louie was a bully. And like all bullies, he thrived on intimidation. He was big and loud, not to mention armed, so no one said anything when he cut another line. As I watched this happen repeatedly, I was glad we were far from the stern and out of his way.

There happened to be a group of four guys between the stern of the boat and where my girlfriend and I were fishing. They seemed like nice, average guys—just out to have a good time on the water and hopefully bring home some bluefish fillets. One of them—shorter and scrappy looking—was quieter than the others, but was quick to smile at his friends' jokes.

Louie's next bluefish was larger; it took him from around the stern and then brought him farther up the bow where he approached the first of the four friends—a tall guy in a white tennis hat. Louie, who at this point may have been drinking, screamed louder than ever, "BETTER GET THAT LINE IN OR I'LL CUT IT!!"

The guy in the tennis hat calmly said: "Hey, relax. I'm reeling in. Don't cut my line."

Louie glared at him, seething with rage. How dare someone stand up to his bullying ways? He cranked in his bluefish as hard as he could, heaved it over the rail, then threw down his rod in a clatter.

Then he screamed inches from the guy's face: "YOU WANNA MAKE SOMETHING OF IT?"

The tall man remained calm and said, "No, I don't want to make something of it, but stop cutting everyone's lines."

Louie turned purple and was about to go apoplectic. Thankfully, the captain, standing on the upper deck above them, intervened.

"Louie, calm down. Go to the back of the boat," he said in a tone of a father scolding his rotten kid . . . again. It was obvious the captain was well aware of Louie's ways.

Louie glared at the guy in the tennis hat for a few more seconds, then grabbed his rod and dragged his bluefish to the stern. The four friends muttered among themselves, shaking their heads. I noticed the small, scrappy guy had a particularly pained expression on his face.

The trip continued on with just a slow pick of bluefish here and there. Louie caught a few more and cut a few more lines, but fortunately remained at the stern.

Finally the captain tooted the horn three times signaling we were heading home. The ride back to port seemed particularly long. My girlfriend and I had caught nothing, not to mention we had been terrorized by a potentially violent lunatic for the better part of six hours. On the way back to port, I promised I'd make it up to her; maybe I'd take her striper fishing next time.

When we arrived at the dock, it was crowded with people. Other boats were already offloading their fares, while others boarded for late afternoon or evening trips. Mates busily scrubbed down decks or hawked their catches to tourists looking to buy fresh fish.

We trudged off the gangplank then made our way through the crowd lugging tackle and an empty cooler. I happened to be walking behind the four friends who were talking about grabbing a beer somewhere. My girlfriend was muttering something about this being her last party boat trip. Just then, we heard a commotion behind us. All of us turned at the same time.

It was Louie. Someone was arguing with him. We could see him through the crowd standing on the dock next to the party boat pointing his finger defiantly at some unknown person and yelling. I sighed. I had seen enough bullying for one day and was about to turn back around when it happened. The smallest of the four friends—the scrappy one—dropped his gear. Then, in the first words I heard him utter all day, he declared: "That's it."

We all watched him walk into the crowd and disappear. Then we saw people moving out of the way of an unseen force headed straight for Louie who all the while continued to argue.

Then, like a periscope locking on a target, a fist rose slowly above the crowd. It got within two feet of Louie, cocked back, and came down mightily and squarely on his chin. Louie instantly dropped between two party boats, which then rose and fell from the wave of a great unseen splash.

There is no more disgusting water than the fetid brine of a summertime boat basin choked with bloated fish carcasses, slicks of engine oil, cigarette butts, beer cans, bilge water, and mats of rotting seaweed. That's what Louie fell into.

The crowd parted again and the short, scrappy fellow emerged with a smile that I will never forget. His friends rushed him, jumping and dancing around in disbelief as he swaggered past. Popeye had just decked Bluto.

But this war was not over.

A minute later, an enraged voice began making its way through the crowd. As it got louder you could see the spray of brine and flecks of seaweed flying.

The voice grew louder still: "WHERE IS HE?! WHERE IS HE?!"

Louie exploded from the mass of people. He dripped with bay water, his eyes red and crazy, a leaf of green seaweed stuck to his bald head.

Then he spotted the four friends and ran up to the scrappy one, the one who—it bears repeating—had just landed one squarely on his chin and sent him headlong into the drink.

"ALRIGHT LET'S GO!!" Louie screamed and began dancing around holding up his dukes like a pugilist from the 1890s.

Enter the police car.

A cruiser screeched up to the scene and two patrolmen got out and restrained Louie. By now a crowd had gathered. People were talking to each other exchanging bits of information of what they had just seen. I suddenly felt honored and privileged to have been one of the few who was there from the beginning, sort of like the proud veteran of some horrible military battle.

One of the patrolmen was now checking each of their IDs. The other was on the radio in the squad car. Louie continued to yell pleading his case to the patrolman. The other guy remained quiet but grinned slightly all the while, further enraging Louie.

And then the patrolman in the squad car got out, turned Louie around, and snapped handcuffs on him. The scrappy guy was politely handed back his ID and given a pleasant "have a nice day" from the other cop. What exactly had just transpired, I will never know for sure, but I believe Louie must have had what they call "priors." Maybe he was on the lam from another fishing fleet where he had wielded his fillet knife one time too many.

What I do know is that Louie was then stuffed into the police car and driven past a crowd of one hundred people who applauded and cheered.

I drove home, my cooler empty and my girlfriend still angry, but with the salty afterglow one gets after a particularly fine day on the water.

The Best Fish Ever

The improbability that this fish was even hooked, let alone landed, has elevated it to this highest of honors. My friend Jim Leedom and I were fishing for Atlantic salmon on Cape Breton Island in late September. It was Jim's first time casting for the "fish of kings," and so far he was having a fine time of it having hooked a fish each of the two days we spent on the Margaree River. The first fish broke off, but the second bulldogged in a deep run then eventually surrendered—a ten-pound colored-up male with a hooked jaw.

Meanwhile I played court jester by hooking water and lots of it while keeping intact a several-year streak of having never hooked a mighty Atlantic.

I had lots of company, it turned out. We spoke with other anglers on the river and learned that success in Atlantic salmon fishing is measured in fuzzy terms like "raising a fish." This means you had the honor of a salmon actually considering your offering before rejecting it. Other anglers who spoke of good fishing, when pressed, admitted they were actually talking about catches that occurred years ago.

By the middle of our last full day, we were getting tired of angler after angler harkening back to days of yore when the river bulged with thirty-five-pounders. Plus neither of us had a touch since Jim's salmon from the day before.

Then we spoke to one more fisherman who agreed that the Margaree, like virtually every other river in the province, might have seen better days. That's when he mentioned the Cheticamp.

He described it as a ruggedly beautiful river that was lightly fished. It flowed out of Cape Breton Highlands Provincial Park about twenty miles to the north. He said you had to hike in to the best pools. Then he closed the deal by mentioning there was a pub right at the entrance of the park—a great place to stop for a beer and a burger after a good day's fishing.

So we drove through the pretty fishing village of Cheticamp and stopped at the visitor center to buy special licenses that allowed us to fish in the park. I remember they weren't cheap, and how it almost seemed like a waste of money seeing as we only had a few hours left to fish.

But we bought them anyway and then began hiking up a trail that paralleled the river. So far the recommendation was spot-on; the Cheticamp was stunning, tumbling through a gorge with some higher forested peaks off in the distance shrouded in low clouds. The rushing water had a deep tannin, almost purplish, stain—like a river full of good brandy.

We walked for half a mile when we ran into another angler coming down the trail on a mountain bike with a rod tube fastened to the frame. He seemed surprised to see us and stopped to chat. When we told him we had just come from the Margaree, he was incredulous.

"Why would you leave the Margaree to fish here?" he asked us, explaining that the better pools were several miles upriver—at least an hour's hike—and that the fishing had been slow for weeks. He rode away, and we began to feel that we had been led on a sort of Atlantic salmon snipe hunt.

With only a couple of hours of daylight left and a steady rain now starting to fall, we decided we should spend our time fishing, not hiking. So we scrambled directly down the bank to the river. There happened to be a deeper run right there—a rarity in a stretch that otherwise appeared to be one long riffle in either direction.

We waded across the river to what looked like the best casting position. Jim generously gave me the tailout, explaining that his two salmon came from similar water on the Margaree. I gladly took him up on his offer and began swinging my Cosseboom through the pool as Jim worked the head.

The strike came almost immediately and a broad salmon went airborne three times in ten seconds. Then it dove for the bottom where it sulked for several minutes before coming up again for three more cartwheeling leaps. After that, the fish seemed to settle down and stayed deep for a long time. Periodically, hard head shakes would thump the rod into dangerous spasms. The tippet was light so the fish needed to be played with extra care.

After a standoff that went on for nearly twenty minutes, the salmon began to tire. It dropped back to the very lip of the run's tailout. Then, before we could get a better angle on the fish, it powered into the riffle and bored downstream.

Now the chase was on with Jim and me slipping over slick, round cobbles while the salmon continued to drop back in the general direction of the Gulf of Saint Lawrence.

Finally, somehow, the salmon stopped. We caught up with it and wound precious inches of backing onto the reel followed by most of the fly line. By now, the salmon was nearly beaten; its fight was down to a few remaining surges and head shakes. Then I waded out and grabbed the thick wrist behind its powerful tail.

Some forty minutes after hooking it and half a mile downstream, a gun-metal gray female Atlantic salmon of about fifteen pounds gasped for oxygen in the river's shallows. Alongside it sat two spent anglers now laughing in disbelief that the fish was really landed. The rain had stopped and the sun was now shining, revealing a beautiful rainbow between two peaks over the highlands in the distance. Angels may have been singing too, but the river drowned them out.

I grabbed the double-hooked General Practitioner wedged in the scissor of the salmon's jaw, pried it free, and flipped it back to Jim, who was too tired to move. I then pointed the salmon into the current and we watched it slowly swim away. Jim reeled up the fly and hooked it to the keeper ring of his rod. My Cosseboom, bone dry as it had been for virtually the entire fight, remained stuck in the cork of my rod.

For this best-fish-ever was not mine; it was Jim's.

Buck Bug Fever

The season after Jim's triumphant catch, I returned to the Cheticamp River. This time, I brought my wife, Mimi, so she could try her luck with the fish of kings. To hedge our bets, we hired a local guide named Robert Chiasson whose services included two extra mountain bikes to get us to the best water. I quickly learned hardcore Cheticamp salmon anglers pedal along this river as much as they cast into it.

After a forty-five-minute trek, we came to a large, cathedral-like pool lined with cliffs on both sides. A lone bike was already there propped against a tree. It belonged to Bob Carpenter, one of the Cheticamp's regulars and a friend of Robert's. We could see him working the deep water beneath the far cliff with booming casts.

We surveyed the rest of the pool. At its head plunged a six-foot waterfall. Robert told us salmon would gather here waiting for high water to scale the falls and reach their spawning grounds. A few large boulders rose from the pool's tailout, and mysterious currents swirled around them.

Robert grew up with the river literally in his backyard and seemed to know every lie that held a taking salmon. He instructed Mimi to stand on a particular boulder along the shoreline. He tied on a Buck Bug—a large bushy dry fly.

Then he pointed and said: "See that triangle-shaped boulder? Drop your fly two inches from the point—not four inches, not six inches. Two inches. Then hold your rod high so the fly wakes across the eddy. There's a salmon there and it's going to come up and eat the fly. When he comes up, let him take it down before you set the hook." There was an air of painful firsthand experience in the last part of his instructions.

Meanwhile, Bob, who had since taken a break from casting, had settled in to watch on the bank behind us. He nodded and repeated the last sentiment in a grave tone: "DO NOT set the hook before the salmon eats the fly."

Mimi nodded. The first two casts were mere inches off the mark. Each time, Robert instructed her to recast. On the third attempt, the bug gently bounced off the rock and landed in the two-inch sweet spot off the point.

"Now wait," Robert said tensely.

Such angling hubris—particularly in extreme-odds fishing like casting a dry fly for a non-feeding salmon—is rarely rewarded. But this was the exception, because three feet of bottom in the shape of a mighty Atlantic salmon slowly rose from the shadow of the boulder. The fish kept rising, revealing cross-shaped spotting across its flanks, hand-sized pectoral fins waving in the current, and a large cotton-white mouth opening wider by the second.

In that remaining instant, we all savored what would happen next. The salmon would take the fly and bring it down slowly. Then Mimi would set the hook—perhaps just after reciting "God Save the Queen" as the British salmon anglers do. Then the hooked salmon would go berserk and leap all over the pool. Robert would eventually land it and photos would be taken of Mimi triumphantly holding her prize before a noble release.

But that's not what happened. For just at that magic moment when the Atlantic salmon's jaws broke the surface of the water and closed around the Buck Bug, Mimi broke. She reared back lurching the fly from the mouth of the great fish, which slowly sounded to the bottom of the pool. We all knew it would not be coming back for a second look.

A collective "Nooooooo!" echoed off the cliffs and down the river as Mimi joined Robert, Bob, and the screaming choir of Cheticamp ghosts all of whom had crumbled at this same sacred altar.

A Bridge Tarpon Too Far

Pssst. Hey you. Yeah, you. Wanna hook a tarpon—a really big tarpon? No problem. Here's what you do: Drive down the Overseas Highway in the Florida Keys and pick a bridge—any bridge. Then fish at night on a moving tide—the faster the better. Cast up-current with a bucktail or a plastic shrimp. Let your lure dead-drift into the shadow line just under the bridge and hang on. You will hook a monster.

The key word is hook, because there is absolutely no way to land them.

My friend Ralph, a former charter boat captain, told me about bridge tarpon. He had just returned from fishing in the Keys, mostly trolling offshore for sailfish and wahoo. When I told him I was heading there myself for a few days but without access to a boat, he rattled off the names of several bridges and gave me some advice on tarpon techniques and lures. It seemed strange; he was so casual about the whole thing as if he was telling me about a local pond that held some bluegills. To me, a place where you could hook a fish that might weigh a hundred pounds was something I would keep from my own mother.

On my first night in the Keys, with a moon rising over the Gulf of Mexico, I headed to one of the bridges Ralph mentioned. I brought the heaviest-action spinning rod I owned rigged with heavy, abrasion-resistant line, an eighty-pound-test shock leader, and half a dozen large barbless bucktails.

Little did I know I was walking into a knife fight armed with spit-balls.

I ventured out mid-span then looked over the rail at the tide rushing toward me. At first it looked dark and lifeless, but then I could make out a distinct shadow line cast by streetlights overhead. The water ran clear with occasional schools of small bait swimming in the current. Then I heard something that sounded like a large piece of concrete had broken off the bridge and crashed into the water somewhere below me. Then I heard another. But it wasn't the bridge; it was the sound of very large fish aggressively feeding.

Looking back now, climbing over the railing and balancing on the edge of a concrete abutment that could fit most of my feet but not my toes was probably not the smartest thing to do—particularly when fifty feet below me, a roaring three-knot current was ready to carry me to Haiti.

But any risk to life and limb vanished when my eyes adjusted to the darkness below the bridge. Just on the edge of the shadow line, half a dozen tarpon were lined up—fish well over five feet long, and weighing up to . . . actually I had no idea how much they weighed since I had never hooked something so large.

The tarpon held effortlessly in the current planing back and forth like salmon riding the steady flow of a river. For a few moments, they looked almost peaceful. Just then, a single ballyhoo frantically skipped underneath the bridge to its inevitable doom. The largest fish broke from the group and ripped it from the surface leaving a crater of whitewater that blossomed for a moment before sweeping away.

With slightly shaking hands, I unhooked the bucktail from the rod's keeper ring. It couldn't be this easy, I thought. The fish must be line shy or ultra-selective about lures or presentation.

They weren't. I flipped the bucktail forty feet up-current then reeled steadily with the tide. Just as the lure swung beneath the bridge and into the shadow line, I saw an immense silver flash followed by the unmistakable thump of a hard strike.

I set the hook and my problems began.

An enormous tarpon jumped so high, with its maw agape and gills rattling, it seemed like it was trying to knock me off my perch and drown me. It crashed back into the water with a piano-sized splash then tore off one hundred yards of line so fast I thought my reel would explode. Somewhere in the darkness beneath the bridge, it leaped again—I could hear the percussion depth-charge splash—then another fifty yards melted off the reel. My rod was now bent in an impossible angle; the butt jutting skyward as I tried to keep the line low so it wouldn't catch on the bridge structure, and the tip pointing down and away as the tarpon continued to take line in long surges. Meanwhile I was stuck facing away from the fish with no room to move.

Then it dawned on me: Ralph never said a word about landing these fish.

My choices were bleak: I could fight the tarpon in this contorted position for the next two hours, where, even if all went well, I would still wind up with a huge, tired fish fifty feet below me. Then what? Or, I could cut my losses right now and intentionally break the beast off. Reluctantly, I chose the latter.

Here was a fish that should have been cast in gold in the memory banks of my personal hall of fame. Instead I found myself snubbing down on the rod and grabbing the reel spool to lock up the drag. When the line popped, the tarpon unleashed one more insane jump that I could only hear somewhere off in the distance.

Over the next hour, there were a few more hookups all ending in the same result: powerful, leaping fish taking me under the bridge and busting up my tackle. Eventually, I gave up.

I have since tortured myself thinking of hare-brained schemes to land these mighty fish. One involves hanging off the abutment by a rope ladder; in another, I leap off the bridge when the tarpon is first hooked then swim to shore with the rod in my teeth to land it. But when I think them through, they always end in the same way: death and/or a lost fish—usually both.

Maybe you can figure it out, now that you know where the fish are and how to hook them.

So have at it. Go nuts.

My Cup Runneth Over

Some old-time striped bass fishermen firmly believe that stripers hate eels. They claim bass only grab them to kill them and spit them out. They will argue you'll never find an eel in a bass's stomach. They say this with the conviction that comes from sorting through a lot of fish guts.

Meanwhile, seemingly every other predatory fish in the ocean relishes eels, from mako sharks to summer flounder. But not striped bass, claim the old-timers, some of whom also believe that stripers ride out nor'easters by eating rocks.

Whether bass love to eat them or just love to kill them, one thing is for certain: A live eel makes one damn, fine live bait for catching big stripers. That is, if you're into that sort of thing.

When I was in my mid-twenties, I was really into that sort of thing. I used eels whenever I could and racked up impressive catches. Eventually, I gave it up like a bad habit. The problem was that I began to feel sorry for the eels. They have such a strange lifecycle: migrating to the middle of the Atlantic to spawn, their young heroically making it back to coastal streams and rivers, sometimes crawling across land to scale dams. It's hard not to admire them. Spending a tide feeding a dozen eels to hungry (or hateful) stripers became increasingly difficult. And yes, you will go through that many on a good night, because stripers can't resist a hooked wriggling eel, particularly when it's tossed into a lively tide rip under the cover of darkness.

One cold November night in my eel-slinging days, I walked to the point of Sandy Hook under a waxing moon and stood among a picket fence of anglers throwing plugs. A school of foot-long menhaden could be seen in the moonlight splashing just beyond the surf-line. Periodically the school would surge to the surface as something large passed underneath.

Sandy Hook Striped Bass

Most of the surf fishermen cast large swimmers to imitate wounded menhaden. But best as I could tell, no one had hooked anything. That's when I reached into the seaweed-lined canvas bag I had slung over my shoulder, pulled out a live eel, and hooked it to my leader. Then I flipped it beyond the breakers and let it sink. Almost immediately I felt a dull thud. I let the fish take a few yards of line then reared back and felt solid weight.

All anglers' brains come equipped with a built-in calculator that immediately tries to determine how big the fish is they've just hooked. Depending on the fight, it rises up and down like the stock market. A long unstoppable run will make the calculator surge higher; a fish floundering limply on the surface causes it to plummet.

Based on this particular fish's initial head shakes—solid but not overly powerful—my calculator stood at ten pounds—a respectable bass, but far from exceptional. Then the fish rolled beyond the first breaker leaving a swirl that looked surprisingly large. The calculator jumped to twenty pounds, even though it hadn't yet fought much.

The next wave dumped the fish right in the wash, where I grabbed the leader and dragged it up on the beach. I shined my headlamp and the fish calculator went haywire. Lying before me was a nearly four-foot-long striped bass—far and away the biggest fish I had ever hooked in the surf. Yet I landed it in under three minutes.

This left me deeply puzzled. For years, I had been hearing surf anglers gush about the heroic fighting qualities of a "cow" striper: stories of hour-long battles in the teeth of a nor'easter; tales of bent hooks and plugs literally snapped in half. Now I had an immense striped bass of my own lying before me that had fought not much harder than the proverbial old boot.

Nevertheless, it was a huge fish, so I found a piece of rope, strung it through its gills, and proudly dragged it past the other surf fishermen to take home and show off some more. Then I laid it in the bed of my pickup truck with my rod, waders, and other gear, and drove off.

I stopped at a gas station on the way home to get my usual five bucks of gas. The attendant walked over, and I readied myself for the inevitable congratulations when he saw my mighty bass laying in the back of the pickup. But he never said a word. I briefly considered stopping at every gas station I passed to fill my tank in one-dollar increments until someone finally noticed my giant fish.

Then I knew who would really like to see a huge striper: my non-fishing parents who lived close by. By now, it was past midnight, but I drove to their house anyway and practically kicked open their bedroom door waking them out of a sound sleep. There I stood, holding the big dead bass.

They were genuinely happy for me, though they hurried me into the kitchen where I wouldn't drip fish slime on the carpet.

My brother woke up too, saw the huge bass, and grabbed his camera. He began taking the requisite glory shots of me standing in front of the kitchen sink holding the fish by the gills. Then it was time to clean the beast. I ran to my truck to get a fillet knife from my surf bag while my brother continued to take close-ups of the fish's impressive head and mouth.

When I began butchering the bass, my parents and brother went to bed, probably not wanting to witness blood and guts while in their pajamas. Cleaning the fish took a long time. I sliced thick slabs of meat from its flanks then skinned them and sealed them in plastic bags. The fish's stomach was empty: no eels inside—or ballast rocks for that matter.

I decided to save the head to make stock for fish chowder. So I got a heavy butcher's knife and started to cut into the backbone to separate it from the carcass.

The knife got stuck on something I couldn't see. There was fish blood everywhere so I just held on and hacked away. Finally the head separated. It was as big as a volleyball with lots of meat still on it. Gallons of good stock would come from it. Then as I rinsed it off under the sink, I spotted a large shiny object deep in its gullet. That's when I looked down its mouth.

Wedged in its throat was a stainless steel cup.

Apparently I had partially cut into the metal while laboring to remove the fish's head, but it was still firmly lodged down the fish's mouth. I pulled hard and eventually removed it. It looked like some sort of measuring cup.

Then it dawned on me—the fish didn't fight because it had mistakenly tried to eat the cup. It made sense: The cup fell off a boat or was dumped at sea. It fluttered in the current looking like a big butterfish or flounder and the bass gobbled it up. I had read stories about cod caught in New England filled with all types of jetsam, including one fish that had a stomach full of light bulbs. And who could forget Quint in the movie *Jaws* telling about the great white shark he once saw eat a rocking chair.

Now I had my own version. The poor bass probably had the cup stuck in its throat for weeks and slowly lost its strength. Maybe I did it a favor by killing it.

The next day, I took the head packed in ice and brought it to a friend who ran a local environmental group. He checked out the head and cup and was suitably astounded. He immediately called a marine biologist he knew and told him to come over right away. The biologist arrived and examined the fish some more, then asked me for details about the catch. I happily obliged regaling him with tales of my angling prowess.

Now this was getting interesting. The big striper may not have fought hard, but maybe I could still get some glory out of this. Maybe my catch would be written up in a scientific paper. Maybe new legislation against dumping garbage at sea would be passed in my honor. The biologist thanked me repeatedly for reporting such an unusual catch.

That night I excitedly called my brother and told him what happened. There was a long silence on the other end of the line.

Then he said: "I put the cup in its mouth. I was trying to find something to prop open its jaws so I could get a good picture. I was wondering what happened to it. Don't tell Mom."

No, I wouldn't tell my mom, but I would have to tell my conservationist friend, who in turn would have to tell the marine biologist what had really happened. They would have a good laugh at my expense. So much for the glory of catching a big striper.

Since then, I have battled large stripers that have left me weak-kneed, living up to their reputation as powerful gamefish. But the question remains why that first big bass didn't fight. Maybe it should have eaten more eels rather than just killing them and spitting them out like the old striper fishermen say.

Skinning Mr. Whiskers

Nothing beats fried catfish. Fillet 'em up, dip 'em in egg, and roll 'em in cornmeal. Deep fry 'em 'til they're crispy. Serve 'em with hush puppies or corn-on-the-cob. Invite all your friends over.

I blame either *Field & Stream* or *Outdoor Life* circa 1983 for putting that idea in my head. One of them published a story about catching giant catfish in the tailrace below a huge southern dam. It was called something like "Battling Mr. Whiskers." In the article, the writer fished live two-pound gizzard shad and loaded up with forty- and fifty-pound blue cats or flatheads—I don't remember which. The article ended with the author arriving home like a conquering hero and half his town showing up the next day for a catfish fry.

My own version of the southern tailrace was a set of deep rapids below a wing dam on the Delaware River in New Jersey. My friend and I started fishing there one spring in high school and found it to be a virtual fishbowl of warm water species. We mostly used bait and caught everything from smallmouth bass, to pumpkinseed sunfish, to eels.

Then one day we got into a school of channel catfish. They were not the monstrous fifty-pounders from the article, more like eighteen- to twenty-inchers that weighed two or three pounds. We quickly learned the best way to get them was to catch a sunny, cut it up, and then drift a chunk through the rapids under a bobber. When the bobber slipped under or tracked sideways, we'd lift our rods and hook another one.

I insisted we keep them, picturing my own big catfish fry for family and friends. This time, I would be the hero and regale them all with stories of my angling prowess between mouthfuls of delicious flaky catfish.

By the end of the day, we wound up keeping a dozen of the largest. We took them home in a big cooler where they stayed alive. Catfish are tough it turns out.

Especially when you clean them. When he dropped me off at home, my friend decided he didn't want his fish, so now I had all twelve to fillet by myself. I laid the first fish on a picnic table in the backyard and considered how I would clean it. Their smooth scale-less skin suddenly felt extremely slippery, and the large spines from its dorsal and pectoral fins stood erect and menacing.

Before I made a mess of the fish and possibly my hands, I called another fishing buddy who had more experience with cleaning game than I did. I explained the situation.

"Oh that's easy," he said. "Get a hammer and nail the fish to a tree."

"Ha ha. Funny," I said. "Seriously, how do you clean these things?"

"I am serious," he said gravely.

Then he went on to explain that I must impale the fish by the head. Then I was to make a long, deep cut just past the gills, grab the skin with a pair of pliers, and yank it off. The skin is very tough, he said, so I would have to pull really hard. When I was done with that, I could take the fish off the nail and proceed to fillet it.

I swallowed and hung up.

Reluctantly, I did as he said with the first catfish. Impaling it while it was still alive was bad, but skinning it with pliers was worse. After that I laid the bloody carcass on the table to separate meat from bone.

When I was done, I had two translucent fillets that measured maybe eight inches long and weighed a couple of ounces each. Cooked, they would constitute a light snack at best. A three-pound catfish is mostly head, it turns out. I looked at my watch; it took me half an hour to clean the first fish. And I had eleven more to go. I looked at them as they swam in the cooler still alive. Tough, hearty fish, those catfish.

I closed the lid and told my mother if she ordered pizza, I would pick it up. There would be no heroics today.

On my way to pick up the pizza, with a cooler full of catfish sloshing around in the trunk, I took a detour to a small stream I knew. It contained a single deep pool where I would sometimes see kids worm fishing for sunfish or suckers.

Little did they know that they, too, would soon battle Mr. Whiskers. I just hope they had the good sense not to try to eat them.

Bottom Fishing

There is an old iron bridge that crosses a quiet stretch of the East Branch of the Delaware River. It once supported mighty Ontario and Western locomotives; nowadays just the occasional farm truck or car rumbles over it. The bridge is painted a traditional sea-foam green, and it contrasts beautifully against the dark, forested mountains beyond it. Below the span, the river gently flows past before quickening into a lively riffle that eventually rounds a bend downstream.

And it was on this bridge I stood gazing at two sleek fish holding in an eddy adjacent to one of its stone abutments.

I had stopped here to scout the river on my way to fish another spot—just a quick check of water levels and to see what might be hatching. It was a pleasant morning in early June—the time of year when mayflies could trickle off at any time. I hadn't expected to see any fish-life except for maybe a few suckers.

So I was surprised when I saw the two fish thirty feet below me gliding trout-like this way and that in the current. They looked greenish-gray like rainbows, though I've seen silvery East Branch browns with that same coloration, too.

Yet I remained unconvinced they were in fact trout since they held in such an obvious and visible spot. A local worm dunker from the hamlet on the other side of the bridge would have long since noticed them and cleaned them out for supper.

So I did what any scientific fisherman would do, which of course was to stand on the bridge and spit on them to see how they would react. When the fish ignored that high-tech experiment, it was time for more rigorous research. I found a pebble the size of a BB and dropped it just above them. Nine out of ten times fish will ignore that, too. Or they will spook, proving they probably were, in fact, trout.

But these fish did neither. Instead, shockingly, they rushed the pebble as soon as it splashed into the river to try to eat it.

The other spot could wait. Here were two nice fish clearly on the feed. I briskly walked back to my car, assembled my fly rod, and slipped on my waders and vest. Then I followed an old fisherman's trail down the steep bank that led to a gravel bar directly beneath the bridge. Glare now made it hard to see into the water, but the seam of current where the trout held was readily visible thirty feet away. Yes, I was now calling them trout, theorizing that the small pebble imitated an egg-laying caddis plunging beneath the surface to deposit its progeny. Indeed a few caddis fluttered above the river giving further credence to my new theory.

I tied on an elk-hair caddis to see if I could coax the trout to take it from the surface. I stripped out some line, false-cast, and then dropped the fly four feet above where the trout held. I let the current take the caddis over the trout, and . . . nothing. So I cast again with the same result. I cast a few more times even twitching the fly over the fish—a technique that often coaxes fussy trout into striking. It did not.

I opened my fly box to try to crack the code. My choice was easy; I picked a bead head caddis nymph that I would plunk into the stream just like the pebble.

I cast again. The line unrolled and the nymph landed in the river just above the eddy with a tiny plop. I watched the line drift a few feet before the leader twitched. I lifted the rod and felt a heavy weight followed by a few satisfying head shakes.

Now I waited for the inevitable hard run or jump—the signature move of a wild Delaware River trout. But it never came. Just more head shakes that now felt slow and lumbering. Oh no, was it? A silvery fish languished on the surface showing large scales and no trout-like spots. Yes, it was a fallfish, aka chub.

Western rivers have whitefish; eastern rivers have fallfish. They are oversize minnows that share the same habitat as trout and frequently fool fly anglers—particularly when they rise to hatching insects. But unlike whitefish, which some anglers target to take home for their smokers, no one deliberately fishes for fallfish. They are flabby fighters, not particularly attractive, and have the peculiar habit of gurgling like a backed-up sink when you unhook them. A fishing friend calls them "failure fish."

The fallfish I hooked—about fourteen inches long—continued its limp struggle. Then, when I had it two feet from my grasp, it did something very odd. It regurgitated a swirling cloud of whitish offal peppered with little flecks of black. What the hell was that? I have seen hooked fish spit up minnows, crayfish, aquatic nymphs, but never anything like this. Yet it looked familiar. What was it?

The answer was above me fluttering and squawking the whole time. A colony of cliff swallows had set up homesteading on the span of the bridge. Their mud and straw nests lined an iron girder running underneath the deck. Periodically, their droppings would splat into the river. Apparently, the fallfish had discovered a convenient, reliable, protein-rich food source: swallow crap.

I landed the fish feeling almost embarrassed for it. In a bug factory like the East Branch of the Delaware with everything from tasty stoneflies to succulent green drakes to feast on, this is the best it could do? But I suppose the fallfish didn't care; food was food.

And of course I threw it back. Because fallfish, as you may have guessed by now, taste like you-know-what.

Poor Grayling

The bush plane bounced down the rough dirt runway, gently lifted off, and then banked southward leaving a dozen of us standing in the western Alaskan wilderness. With a light rain falling, the guides immediately went to work inflating sturdy aluminum-framed rafts. The rest of us tugged on waders and rain gear from waterproof duffels and strung up fly rods.

One of the guides told us it would take an hour before they were ready to start the float—a week-long, 150-mile odyssey for salmon and rainbow trout. He said we could fish if we wanted here at the put-in, but not to stray too far.

So I wandered upriver a few hundred yards. This was a headwater stream, glass-clear and rocky and trout-like, maybe fifty feet wide and lined with scrubby alders. I eventually came to a slower, deep pool with a bubbly tongue of current flowing in at its head.

I rigged a trout leader and tied on the classic Alaskan dry fly: a size 14 Mosquito. I stripped a few yards of line from the reel, made a false-cast, and dropped the fly in a slick behind a submerged boulder. It drifted for a few feet among the bubbles. I tensed with the feeling it wouldn't get much farther. It didn't. A gray shape rose off the bottom and took the fly in a confident rise.

I lifted the rod and a fifteen-inch pewter fish jumped almost nonchalantly before boring for the bottom—a grayling. It gave a few more mini runs and head shakes before allowing me to eventually slide it into the shallows. What it lacked in brute strength, it made up in exotic good looks: turquoise and salmon-colored flecks covered its outsized dorsal, making it seem more suited for a coral reef than the harsh realm of the sub-Arctic.

I let the fish go and watched it swim slowly toward the center of the pool. I readied for another cast stripping off a few arm lengths of fly line while simultaneously roll casting to straighten the leader. On the last roll cast, just before I was about to shoot the line back to the head of the run, the fly happened to touch down just ahead of the released grayling still making its way to the center of the pool. And as if the last five minutes had never happened, the fish immediately swam over to the Mosquito—and ate it for a second time.

Like my own version of Groundhog Day, I set the hook—again—and landed the grayling—again. Not surprisingly, the fish came to my hand quickly. It felt strange releasing it a second time. All this high-tech tackle—the high modulus graphite rod, invisible fluorocarbon leader, chemically sharpened hooks, and Gore-Tex waders—used to catch something so naive it could be caught and released two times in as many minutes. It seemed unfair.

But the grayling would have the last laugh. By the second day of the trip, after the aforementioned light rain turned into a thirty-six-hour deluge, the river jumped three feet and ran the color of chocolate milk for the remainder of the trip. The grayling remained safe—as did the salmon and rainbow trout—and all my high-tech tackle was rendered useless.

The King of the River

Once I hooked the king of the river. It was a warm afternoon in mid-May and I was casting for shad on the Delaware. All of the fishermen around me were hooking shad. Some of them hooked large roe shad up to six pounds. They ran and jumped and tail-walked like baby tarpon.

I wanted to hook a big roe shad; but instead I hooked the king. The king ran slow and long and ponderous. Then the king stopped and held in the current at an angle where I couldn't turn it.

Minutes passed. While I stood unable to budge the king, some anglers hooked, fought, landed, and released six-pound roe shad, then cast and hooked up a second time.

The king fought for so long, anglers eventually stopped fishing and began to ask me what I had on the other end of the line. I didn't know I had hooked the king, so I simply said: "I don't know."

Some thought it may have been a muskellunge. At one time the state record came from the Delaware. It was caught on a gold shad dart and weighed forty pounds. The muskie must have thought the dart was a shiny tidbit worth eating.

I was using a gold shad dart.

Jim and Aaron caught a muskie, too, not five miles from where I battled the king. They were using live minnows trying to hook two-pound smallmouth. Instead they hooked a twenty-five-pound muskie.

Maybe I had hooked a muskie.

My late friend Dave Palmer once told me he saw someone land an Atlantic sturgeon just downstream from where I stood. This was in the early 1950s. He said it was six or seven feet long.

Maybe I had hooked an Atlantic sturgeon.

The king continued to battle. I turned and looked upriver and downriver. Ten anglers I didn't know had stopped fishing and were watching on either side of me, the butts of their rods resting on their hips like idle shotguns.

Now the king made its move and began to swim toward me. It was finally tiring twenty minutes after I hooked it. As it swam closer it began to push a bow wave. Ten fishermen I didn't know strained to see it through the glare.

Then the king of the river swam into full view. It shined gold the way a king should. There was a collective groan by the ten fishermen who immediately walked away from me and began casting again for six-pound roe shad. They disrespected the king.

It would be another ten minutes before I eventually landed it. I had to wade into the river and maneuver the king between me and the shoreline. Then I literally chased it down where it beached itself in the grassy shallows.

I twisted the gold dart from its tough rubbery lip while it thrashed and splashed me with muddy water. Then the king swam away looking inconvenienced but little more.

Yes, the king of the river was a twenty-pound carp. For twenty-five years and counting, it remains the largest fish I have ever hooked in the Delaware River. I cannot overthrow it.

Hail to the king.

Part III

STREAMSIDE HAZARDS

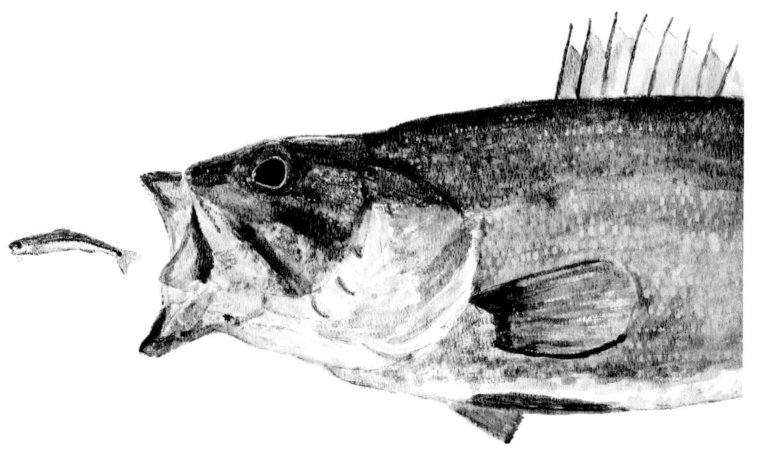

Loon Lake Smallmouth

Fishing with Griz

The first thing you need to know when you go fishing in Alaska is that you are no longer at the top of the food chain—you are now in the domain of Griz.

The second thing is that "Griz" is what most Alaskans call grizzly bears. Not "a griz" or "the griz." This is not the place for indefinite articles or other grammatical doo-dads. Just plain old Griz.

And the third thing you need to know is that Griz can eat you.

Ten of us were scattered along an exposed gravel bar along Alaganik Slough—a tidal creek in Alaska's vast Copper River Delta near the fishing village of Cordova. The group consisted mostly of fishing friends and acquaintances, along with two older Alaskan gentlemen who had puttered up to the spot by cartop boat. Our catches of gutted and bled silver salmon lay scattered around us. More recently landed fish still flopped around.

A tannin-stained tributary poured into the glacial-colored water alongside the gravel bar. Salmon staged briefly here before heading farther upstream. Here, a well-cast pink pollywog—a fuchsia feather-duster of a dry fly made of spun deer hair and marabou—would be chased down and engulfed by the non-feeding silvers for reasons known only by their own kind.

The salmon came in waves. Sometimes, every rod would be bent. Then half an hour would go by without so much as a swirl.

During one of these lulls I wandered away from the group to fish the slough above the creek. I saw the wakes of a few salmon cruising here and there, so I followed them along the bank into a deepening thicket of willows.

Then it occurred to me: This seemed like a really good place to be mauled and eaten by Griz.

I had already learned about "bear etiquette" from various guidebooks and knew the cardinal rule of never surprising a grizzly bear. The easiest way to avoid this is to talk loudly whenever you are in bear country. This gives bears the opportunity to quietly retreat. Apparently, despite fearsome reputation, Griz is a shy, retiring creature. That is until it wants to eat you.

So to ward off Griz, I gave a loud, stream-of-consciousness babble as I entered the thicket: "NOW I'M WALKING ALONG THE BANK AND I THOUGHT I SAW A SALMON. THINK I'LL SWITCH TO AN EGG-SUCKING LEECH. BETTER RE-TIE MY LEADER." I did as I announced, and began casting the fly out and stripping it back with sharp pulls. Just before I picked up the line for another cast, the fly disappeared in the hooked maw of a fifteen-pound silver that proceeded to jackknife and run all over the slough.

I released the fish and decided to rejoin the group. "THAT WAS A REALLY NICE FISH, BUT I THINK I'LL LEAVE NOW AND SEE HOW MY FRIENDS ARE DOING. . . ."

Five minutes after I got back, I looked upriver to where I was just fishing and thought, "I don't remember seeing a brown Volkswagen parked up there." It was a four-hundred-pound grizzly bear standing along the bank in the exact spot where I landed the salmon. It may have been hiding in the willow thicket the whole time.

Everyone stopped fishing to look. For some, it was their first time ever seeing a bear. My friend Rob immediately rummaged through his backpack and produced a video camera.

Then Griz began a pigeon-toed saunter toward us.

This gave some of our group the opportunity to offer their recently acquired grizzly bear knowledge.

Dave went first: "I read that a grizzly bear won't approach a group of three or more."

We all nodded in firm agreement.

But Griz kept coming.

By now, I noticed that the two elderly Alaskans had quietly gathered their salmon, loaded their gear in their launch, and began rowing to the safety of the far bank.

John's turn: "We're upwind. I read that if a bear smells you they'll retreat."

A couple of us still nodded once or twice.

Griz kept coming.

By now I found a long piece of driftwood, rammed it through the gills of the three salmon I kept, and slung it over my shoulder like a caveman. I planned on taking it with me, but would drop the bloody mess behind at a moment's notice if necessary.

Griz kept coming.

Everyone else quickly gathered their salmon and tackle and began to head downstream to the path that eventually led back to our cars.

That is, everyone except Rob. A former *Newsweek* photographer who had been through civil wars and riots, he seemed to almost relish the looming danger as he trained the camera's viewfinder on Griz, who, of course, kept coming.

We pleaded with Rob to join us, but finally gave up announcing that we really, truly were leaving and that he was on his own. And we did, heading around a point where we could no longer see him or the bear.

But no sooner had we rounded the bend when Rob came running to catch up with us.

"Wow that bear kept coming!" he said, panting with an air of adrenaline-fueled exhilaration.

Rob told us the bear never stopped; it walked right to the gravel bar and began scrounging at the offal left behind from the gutted salmon.

Then he said, almost as an afterthought, that the bear was still heading our way.

Just then, my friend Jim showed up from downriver. An hour earlier, he had decided to fish another spot and had missed the whole encounter. I met Jim in 1990 waist deep in the Delaware River fishing for shad and we have been friends ever since. He is intense and excitable. When we told him there was a grizzly coming down the slough, he got a crazed expression on his face.

"Where? *Where*?" he demanded. "I gotta see this!"

Now all of us, including Rob, pleaded with him, warning that the bear was just around the bend, and that it was too dangerous. But there was no persuading him.

With camera in hand, Jim walked purposefully around the point. The group waited. Less than a minute later he came stumbling back covered in mud and babbling. We couldn't quite understand him: "Bear twenty feet away . . . looked right at me . . . head like giant pumpkin . . . tried to run . . . slipped in mud . . . couldn't get up . . . crawled back."

He collected himself, took a deep breath, then loudly proclaimed, "GOD that bear is close."

We all hurried back to the safety of our cars and called it a day.

Later that evening, in a fishermen's pub, we wondered why this bear had ignored every canon of bear safety written in the literature. Why did it keep heading toward us? Didn't it see that we were a group of ten boisterous humans? Didn't it smell us? Aren't they supposed to be shy and retreating?

The answer was obvious: Griz doesn't read.

The Legend of Pusfoot

Don't call me Ishmael; call me Pusfoot. That is, at least for one week each spring on the Upper Delaware River when a certain group of sturdy male adventurers gather for a canoe and camping trip.

It was on this trip more than two decades ago when I had the misfortune one night of trying to warm my feet against the wrong rock at the campfire. This caused a chain reaction: the rock moved, the grill over the fire shifted, and a pot of boiling water poured into my right shoe. Searing pain ensued, followed by a mad scramble to the nearest cooler where I plunged my foot into melting ice among bottles of beer.

But the damage was done—a second-degree burn on the top of my foot about the size of a doughnut that throbbed with pain. A quick survey revealed that no one had brought a first-aid kit except for a tube of some sort of cream that was used for either insect bites or jock itch. Whatever it was, half the tube was emptied on my foot, a sock was put on as a makeshift bandage, and I was handed another beer.

By the next morning, after the wound had spent most of the night oozing, a hard, yellowish crust had formed bonding my sock to my foot as sort of a de facto scab. I showed this strange medical phenomenon to the group for their opinion. As one might imagine, ten guys on a camping trip can be an extremely sensitive and caring lot. This group was no exception. Jokes about gangrene and amputation rained down, with the occasional zinger thrown in about bloodlettings and leeches.

Then one of them said: "Looks like someone's got a new nickname: PUSFOOT!" This was followed by roars of laughter with everyone repeating the name a few times as practice for the coming years.

And so the name has stuck like my sock did for those next several days—until a doctor cut it away with surgical scissors after the trip ended, treated me with antibiotics, and threw in a tetanus shot for good measure.

Peas Eddy Brown Trout

But what I remember most from that trip were the green drakes.

It was my first serious attempt at fly fishing for wild trout. The season before, on a fly rod I could barely cast, I had lucked into a large rainbow that immediately broke me off. That fish haunted me, and I spent the offseason thinking about it.

The day before the camping trip, I stopped at a fly shop to buy some tackle and ask advice. The shop owner picked out a dozen flies, mostly caddis, sulfurs, and March Browns. Then, in a hushed tone that bordered on reverence, he said, "Maybe you'll see some green drakes," adding a few gigantic chartreuse dry flies that bristled like old-fashioned shaving brushes.

I nodded and said, "Yeah, I hope so," having no idea what a green drake was.

Now, just forty-eight hours after his bold prediction, I found myself gently easing on my waders so as not to rub them against the burn. Then I slid my canoe into the river with my fly rod draped across the gunnels.

The rest of the group stayed behind content to eat and drink by the warmth of the campfire.

I pushed off and began paddling to the far side of the river. Midway across, someone called out, "Good luck, Pusfoot," followed by more guffaws.

I beached the canoe and found a spot in the tall riverside grass to sit. As evening set in, hatches of insects started coming off the river like acts in a play. First came the caddis—hyperactive and moth-like bouncing and skipping and swirling over the water like wind-whipped snow flurries. Next, as the light continued to dim, sulfur duns floated downstream, miniature regattas of delicate yellow insects riding stoically in the current.

All the while, fish rose sporadically. I struggled to cast to them with no luck.

It was nearly dark when I saw a handful of much larger bugs riding down the river. They looked whitish green and were many times the size of the biggest sulfur. I realized they must be green drakes.

More and more appeared in the waning light, and trout began to appear seemingly out of nowhere to rise to them. Their gulps could be heard over the river's gentle purl. I clipped off my sulfur and tied on one of the giant cartoonish flies I bought in the fly shop.

I made a graceless cast, the bulky fly making a distinct swishing sound. When it landed, it suddenly didn't seem so big, and in fact, resembled the hundreds of flies now floating all over the river.

It was nearly dark. The river had faded into little more than a silvery mirror. A trout began rising thirty feet from me, eagerly taking the drakes and leaving heavy black rings that rolled downriver and broke up in the current.

I tried timing my next cast to land just above where the fish held. I picked the fly off the water, false-cast, and shot the line. But I misjudged the distance and dropped the fly two feet past the fish and slightly downstream.

But it didn't matter because less than a second later, the trout turned and sucked down my fly in a confident slurp. I lifted the rod firmly. It jerked into a deep bend as fly line hissed off the river and flew through the guides.

The trout leaped then rushed downstream spinning the handle of my Pflueger Medalist. I chased it joyfully, stumbling through river grass and listening to the drag ratcheting away. It was a scene I had played in my head for the past year; now I relished every moment.

Eventually, I coaxed the fish into slack water, and a minute later it laid in the shallows—a heavy brown just under eighteen inches, the first wild trout I ever landed on a dry fly. I eased the waterlogged green drake out of the fish's jaw and let it go with trembling hands.

It was ten o'clock. I wandered back to the canoe in a contented daze. An owl flew silently overhead, its rounded wings silhouetted against the last slivers of dusk.

Across the river, the campsite glowed. I could hear laughter and good cheer coming from my friends who would want to hear the story about the big trout and how I landed it.

I paddled back oblivious to any pain that may have still lingered in my foot.

All of these years later, you can still see a faded scar that commemorates not a clumsy moment around a campfire, but this greatest of all days.

Yes, you can call me Pusfoot.

Tackle-Busters

When my son was six years old, he hooked a nice bass in a weedy lake at my father-in-law's cottage in southern Canada. We were drifting through a channel when the fish hit, pulling the rod so hard it almost went overboard. "I've got a big one," he yelled, trying to crank against the drag that yielded line in choppy spurts. This was his first big bass, hooked on a new spinning outfit I bought him especially for the trip.

Then the fish bored into a mass of weeds and became an immovable weight—too much for the four-foot fisherman on the other end who was now looking at me with a helpless expression on his young face. Time for Dad to save the day. I gave him the reassuring nod of a father who was about to demonstrate how an experienced angler handles this sort of thing. Then I took the rod, gave it one good heave, and snapped it in half. My son immediately burst into tears.

Somehow, the line didn't break, and we managed to eventually land the fish—about a two-and-a-half pounder. There's a picture somewhere of the catch, my son half-smiling through tear-streaked cheeks.

On the way back to the dock to make him feel better, I told him how all seasoned anglers have broken a rod one time or another.

"Really," he said. "Have you ever broken your rod?"

I sighed and told him yes, many times.

So for the rest of our vacation, he insisted I tell him about every rod I ever broke. And so I did:

The first time it happened, I was sixteen years old and riding my bike to a local lake. I had permanently borrowed my brother's rod—a light fiberglass six-and-a-half footer that originally came in a blister pack. He had taken it saltwater fishing, and the cheap reel that came with it had long since corroded into a rusty artifact. Since, like all of

Loon Lake Largemouth Bass

my family, he didn't really fish, I requisitioned the rod and fit it with my favorite surf reel. This was long before I knew of such angling axioms as "well-balanced tackle." All I knew was that the bulky reel could technically be screwed onto the reel seat and that was good enough for me. And the entire outfit draped nicely over my bike's handlebars. Who cares if the reel was four times as heavy as its predecessor?

It turns out the laws of physics cared. So did the law of gravity.

When I hit a pothole that rattled my bike, I suddenly noticed that the reel seat with the giant surf reel attached was now dangling from the rest of the rod at a weird angle. On closer inspection, I discovered that the fiberglass had collapsed at the butt like a folded cardboard tube.

Oddly, when I took the weight of the massive reel off the rod, it sprung back into place, making me think that it had somehow miraculously fixed itself. Wrong. When I got to the lake, every time I made a cast, the rod failed again, making the sickening sound of broken fiberglass rubbing together. At least I salvaged the tip to build an ice-fishing rod.

A year or two later, I broke a lovely, expensive new spinning rod made of boron, which at the time was considered the next quantum leap in rod technology. Light and extremely sensitive, boron had it all, we were told. One rod manufacturer boasted that you could touch a feather against the rod tip and feel it right down to the reel seat. Imagine how a hard-hitting bass or pike would feel! But boron had one drawback: It was brittle. Just how brittle, I learned one day as I fondled the rod at home.

All anglers are occasionally guilty of fishing in the living room, particularly with new tackle, and I am no exception. I flexed the rod this way and that imagining how much it would bend as I wrestled a big bass out of heavy cover. I flexed it a little more. I'll bet a big bass would bend the rod about this much . . . or maybe even more like this . . . SNAP.

A beautiful two-piece rod had now become a not-so-beautiful three-piecer. Physics again, along with the Periodic Table of Elements, conspired against me.

Then there was a rod-breaking incident that remains cloaked in mystery. Many years ago, I found a very nice graphite spinning rod listed for a surprisingly reasonable price at a high-end sporting goods store. At the time, a decent graphite rod was at least eighty dollars; this cost less than half that. It was a brand I never heard of and whose name now escapes me. But I remember the model clearly: It was called a "Lucky Stik" and had a little shamrock next to the name.

I inquired with the head of the fishing department why the rod was so inexpensive.

He examined the rod carefully, waving it in the air a few times, inspecting the wrappings on the guides, and checking out the reel seat. He looked at the price tag again, raised his eyebrows, and handed the rod back to me with a puzzled expression. "I have no idea," was all he said.

Before he could look it up to see if it was mismarked, I ran to the cashier. Sold.

What adventures Lucky Stik and I had together! Bass, trout, crappie, catfish. These were days of joyful spin fishing between college classes, and the Lucky Stik had become my trusted companion, like a cowboy's favorite six-shooter.

Then I left it in my girlfriend's car—a girlfriend who on several previous occasions had accused me of paying more attention to fish than to her.

Now I'm not saying for a moment that the rod was broken on purpose. All I know is that the next time I saw the Lucky Stik, twelve inches of the tip were gone with no plausible explanation as to what happened. Shortly thereafter, she relieved me of my boyfriend duties entirely. Let's move on.

I've broken three fly rods by simply rigging them up, pulling leaders through the guides, and having the bulky leader knot catch on one of the top guides snapping off a foot-and-a-half from the top. Why tackle manufacturers haven't devised a way to streamline this connection so it slips through the guides is beyond me. It's almost as if they want you to break their rods so you'll have to buy more. Hey, wait a second. . . .

Car doors are notorious destroyers of fine fishing tackle. I can think of one example that wound up being a two-for-one deal, though it wasn't technically a door. I was going surf fishing with a friend and had borrowed my parents' car, which boasted electric windows—at that time quite a luxury. The rods didn't fit in the car all the way, so we fed the top ends out the passenger-side window left half open.

We headed south on the Garden State Parkway excited about chasing bluefish or possibly stripers in the surf. I had just paid one of the parkway's infamous tolls, and started to roll up the driver's window with the nifty push of a button. But nothing happened, so I tried again. Then my friend screamed, but it was too late. I was pressing the wrong button—the one for the passenger window. Yes, I snapped the tips off both rods cleanly. I don't remember if we went fishing anyway, but I doubt it.

I eventually had my rod rewrapped into a very nice heavy jigging rod that was then stolen from a car parked in a questionable location in New York City. There was definitely bad karma in that rod that I hope was passed on to the next owner.

I can think of three rods I actually broke while fighting a fish. Two were fly rods broken a day apart from each other in Alaska while battling silver salmon. The forty-ninth state seems to have cornered the market in terms of rod breakage based on straw polls among other anglers I know who have fished there. My theory is that hooking fish after fish as one can do in Alaska makes the angler careless, and perhaps impatient, eager to get the fish to the net just so they can cast again to hook another. I'm as guilty as the next person.

The other broken rod occurred in Florida while wade fishing a tidal flat. I was blind casting a bucktail jig and hooked about an eight-pound redfish that instead of running and bulldogging away, swam right at me. I reeled like a madman trying to maintain tension on the fish, which kept coming. I held the rod higher and higher still until the fish swam by maybe five feet from where I stood. The angle was too much for the little pack rod I was using. The top two feet broke like a toothpick then slid down the line as the fish passed me and kept going.

Like my son, I managed to land the fish despite the broken rod, except I didn't cry—at least not on the outside.

The most recent incident took place in Canada again, in fact on the same lake and nearly the identical scenario as when I broke my son's rod.

This time it was a cherished twenty-year-old vintage Lamiglass graphite spinning rod with a light tip but surprisingly powerful action that you just don't find anymore. Over the years, the rod had taken countless bass, shad, walleye, and even some stripers and snook. It met its demise when a lure caught in a submerged weedbed on the windward side of the boat. We proceeded to drift over the snag and kept going, forcing the rod into an impossible angle. It broke cleanly.

Except this time it was my son holding Dad's rod, not the other way around. Revenge? Another broken-rod mystery? Or just another entry to an ever-growing list.

Hooked on Hooks

Ten yards off the limestone breakwater, fish boiled on schools of small bait. The plug touched down, traveled less than a foot, and was walloped. A big snook went airborne in an end-over-end leap. It crashed back in, took off on a long run, and then jumped again. The angler scrambled along the breakwater trying to keep the fish in front of him while pumping it closer.

It leaped one more time before tiring. The angler brought the fish within reach, grabbed its bass-like mouth, and hauled it out of the water. He put the rod down and started to remove the big 3/0 treble hook from its jaw. But the snook, in one last burst of energy, coiled its body then unleashed a mighty head shake yanking itself free from the fisherman's grasp. It fell, and was about to hit sharp limestone where it surely would have injured itself badly if not fatally.

But fortunately, the snook was saved—not by some graceful catch by the angler, but by the rear 3/0 treble hook. It plunged deep into his angler's wrist and was now embedded past the barb with the twelve-pound fish dangling relatively safely from the other end.

The angler was me.

To say I screamed like a little girl would be an insult to little girls. A piercing shriek echoed across the Gulf of Mexico sending a dozen roosting pelicans to the air. My friend fishing fifty yards away came clambering along the breakwater to see what happened. He found me lying there literally hugging the snook. I begged him to get the fish off the other treble hook before it started thrashing again.

He did and got the fish back in the water. But the damage was done. A nine-inch plug was now a permanent part of my anatomy. Adding insult to injury, snook continued to blitz schools of bait just an underhand cast away.

He was a good friend. Rather than keep fishing, he took me to an emergency room where a doctor numbed my wrist and pushed the hook through before cutting it off with heavy-duty pliers. It turned out he was a snook fisherman himself, and he gently pumped me for information, an interrogator complete with instruments of torture. I sang like a canary.

The next morning, with a heavily bandaged wrist and a fresh tetanus shot, I returned to fish the same spot. Everything looked the same, except the snook were now chasing baitfish two hundred yards off the breakwater well out of casting range. Still plenty were caught—not by us, but the doctor whom I immediately recognized. He was anchored in his boat fishing live bait and getting snook on every cast. He waved to me between landing another fish. I waved back to the son-of-a-bitch.

Hooking oneself ranks as one of fishing's major occupational hazards. It is not nearly as common as sunburn and insect bites, but far more prevalent than, let's say, falling out of your boat or getting your finger chomped by a pike or big bluefish.

The good news—if there is any—is that not all hooks that accidentally wind up in your anatomy require a doctor's visit. For the do-it-yourselfer, some can be removed using a length of stout fishing line. You wrap the line around the hook's bend while pushing the eye downward into your skin. This stretches the entry point just enough so the barb can exit cleanly. Then you give a sharp tug on the line and snatch out the hook.

The only catch is that you need two hands to do this—one to press down on the hook's eye, the other to yank the stout line. This is fine if you hook yourself in the leg, but can be problematic if you hook yourself in other places. I found this out the hard way while trout fishing alone on the Upper Delaware one spring. A sloppy backcast aided by a light breeze slapped the line and leader across my upper arm digging the fly in just above my elbow. I waded ashore and tried to figure out a way to use a tree branch as another set of hands—sort of like the old doorknob trick to extract a tooth. Luckily, before I injured myself any worse, two well-dressed fly fishermen happened upon me. They showed pity and popped out the hook, but not before lecturing me on how barbless flies do less damage—to trout, not me.

This technique wouldn't have worked for my friend Rob, who wound up with a heavy gauge salmon fly in his neck during a group fishing trip to Alaska. The salmon were in heavy that morning with silvers hitting on almost every swing. Then Rob hooked himself. The excellent fishing must have clouded our judgment, because we actually contemplated doing the stout-line trick. Something about the angle and the location of the hook made us reconsider. Fortunately, this happened just outside of a small fishing village with a local medical center, which we wound up visiting. After the doctor removed the fly, he assured us that waiting was the right thing to do. The hook turned out to be millimeters from Rob's jugular vein. Yanking it out might have proved . . . messy.

The last time I hooked myself past the barb turned out to be the last time I used barbed hooks. It happened one night in the late fall on what turned out to be the best night of surf fishing that season. A school of keeper-sized striped bass held in a tide rip and slammed plugs on every other cast.

I had just dragged another ten-pounder on the beach. It was late—well past midnight—and a stiff wind blew out of the northwest. As I reached into the fish's mouth, I could feel that the cold had robbed some of the dexterity from my fingers. I grabbed the treble hook anyway, just as the bass began thrashing. A tug-of-war broke out between the fish and me, and I could feel myself losing. And I did. The hook wound up buried next to the cuticle of my thumb. Then, for good measure, the bass continued to thrash as I bled all over the beach. Luckily, a friend was nearby and he kneeled on the fish as he worked the hook out of its jaw.

He separated the bass from my thumb. Then he asked me what I wanted to do with the fish—keep it or let it go. We had been releasing bass all night, though I had made allusions earlier that night to taking one home to make bass chowder.

"Kill it," I said before he could finish his sentence. I am a sore loser.

Afterward, I wound up in an emergency room getting the hook yanked out for about what a new rod and reel costs, with a couple of nice wooden plugs thrown in. I was young and had no health insurance.

The next day, I took a heavy file and a pair of needle-nosed pliers and flattened down the barb of every lure and fly I owned.

Fortunately, my own experiences have been largely private matters shared by just a few fellow anglers who happened to be with me at the time. This was not the case for a fisherman in a crowded emergency room at the New Jersey shore one Saturday.

My friend was there with his buddy who had bruised his shoulder in a waterskiing accident. The emergency room was filled with other patients nursing various injuries and illnesses.

As my friend waited, he heard raised voices and saw a man at the nurse's station arguing. The man's back was to the rest of the waiting room, but a heated discussion could be heard clearly. He was pleading how he had been there a long time and demanded to know when he could see a doctor. The nurse calmly explained that there were many other patients before him, and he would have to wait his turn.

My friend turned to his buddy and remarked how selfish this guy was—that he looked perfectly fine and was probably there for something really superficial like a bruised pinky.

Then the man turned around. An enormous wooden striper plug hung embedded in his cheek. He sat back down with the entire emergency room staring at him probably feeling much better about their own respective maladies.

Losing the Bunker Cup

Years from now, someone cleaning out their attic will find a ridiculous lime-green fishing lure the size of a hubcap with dozens of names and dates and fish statistics scribbled on it. They have found the Bunker Cup—the coveted trophy from a fishing contest started on Block Island among a dozen fishermen who rent a house there every September. Don't worry if you never heard of it; it is not on the BassMasters tournament trail. It has no corporate sponsors, though we would certainly entertain offers if you know anyone.

The contest always takes place on the Friday before the end of our week-long stay. Here are the rules: The biggest fish, regardless of species and method-of-capture, wins. Methods can include—but are not limited to—surf fishing, boat fishing, fly fishing, bottom fishing, still fishing, and even spear fishing. The last method has caused ire among some of the participants who state that the spear fishermen in our group have a distinct advantage. Others point out that bad weather can keep the spear guys at the dock and out of the hunt.

I am proud to have stayed completely neutral on this issue of spear fishing. So let's change the subject completely and talk about the past winners. Top fish have ranged from a thirty-five-pound striped bass shot by one of the miserable, cheating, spear fishing poachers in our group, to a ten-pound bluefish honorably caught by one of our fine, skilled surf fishing sportsmen.

The winner gets to inscribe his name on the Bunker Cup trophy, along with the date, and the type and size of fish caught. This is done with a black Sharpie. Then they receive the highest privilege of all: displaying the Bunker Cup at home for the entire year. This is a rare honor since the trophy is a one-pound bunker spoon roughly the color and size of a leprechaun. It has a blood-red keel, a fixed 10/0 hook that looks as

Block Island Northern Sennet

though it could land a great white, and a 6/0 treble stinger hook dangling from its end. Rumor has it some have lost on purpose because this weighty privilege is too much to uphold.

Others have become quite competitive about winning the Bunker Cup, risking life and limb—literally—as I once found out firsthand.

One year I got Bunker Cup fever. I had never won before, and had decided this was the year to take home the trophy. Winning would be like wearing the Green Jacket at the Masters, except it would be much better because it would have nothing to do with golf. And I had a perfect spot on my mantel where it would reside. No, my wife did not know this yet.

I decided to begin fishing at 12:01 Friday morning. Paul and I ventured onto a tidal flat where I knew striped bass would be herding baitfish under the cover of darkness. I brought an eight-weight fly rod; Paul used a light plugging outfit and some small swimmers. Sure enough, we soon heard the distinct *swishhhhhhh* of showering silversides followed by the loud crash of gulping stripers. We excitedly cast. But it wound up being too much of a good thing with fleeing bait everywhere, yet only a few very selective bass scattered on the flat. Despite a few decent shots at breaking fish, they would not take our offerings in the shallow, clear water.

Eventually the tide ebbed and we left, stumbling back to the house at 4:00 a.m. In an hour or so, the other members of the group would begin the morning shift of chasing The Cup. The good news was that an offshore tropical storm pushed a heavy swell that would keep the filthy, unsportsmanlike, cheating spear fishermen stuck at the dock. But it also churned the surf into a roiled, foamy chocolate egg cream.

This restricted us to the limited protected waters on the island. Some of the group fished the inlet at the Coast Guard Channel for a shot at the island's mercurial false albacore that sometimes tear past. I eventually caught up with Jim and Aaron at the docks at Old Harbor in the afternoon. They were jigging up small baitfish and then live-lining them hoping for larger game. In years past, this has paid off with keeper fluke, bluefish, and striped bass. Though so far, they had caught nothing.

I fished with them, but began looking at the high breakwater jetty that protects the southern end of the harbor. I knew the swell from the tropical storm was creating a major whitewater rip at the end of the breakwater—the perfect place to hook a nice striper, the kind of fish that would take the Bunker Cup.

But there was a catch. I had to get to the end of the two-hundred-yard-long jetty. No easy task since huge waves periodically exploded along a fifty-yard stretch where the breakwater curved halfway out. To get past that section, I would have to time the waves just right.

As I watched the waves, I noticed the swell seemed to be subsiding. Plus the tide was dropping. Yes, it was definitely getting safer out there, I kept telling myself. Clearly I had the fever—bad—for the next thing I knew, I had put on my full surf gear—waders, belt, studded cleats, foul-weather top, plug bag, and nine-foot surf rod.

Then I began to pick my way along the jetty rocks. But I stopped before I approached the section receiving the brunt of the biggest waves and began to think better of it. Even though the surf did seem to be calming down, there were still enough big breakers to make reaching the end downright dangerous. Too risky, I thought; better wait here and let the tide drop some more. Then I turned my back to the breakers and began casting to the quiet protected water on the other side of the jetty.

Perhaps the freak wave originated in the Azores. It made its way west across the Atlantic; then as it neared the Rhode Island coast it gained energy in the low pressure from the tropical storm. By now, it was a relative monster. As it approached the shore it heaved upward readying for a final explosion of glory like a kamikaze pilot about to take out anything in its way.

And that would be me—a flea about to be swatted by the hand of Poseidon himself.

I had just made another cast. After that, all I remember is hearing a sudden very, very loud crash behind me instantly followed by a wall of water shoving me with a terrifying force I hope to never feel again.

I fell six feet and landed hard in the jetty rocks as thousands of gallons of seawater poured over me. The wave subsided, and I lay facedown and crumpled for a few moments in shock at what just happened. I needed to get out of there before another wave came. I pulled myself up and crawled to the top of the jetty then stumbled back toward the dock.

I must have been a pathetic sight. Half my plug bag had spilled out during the fall and now a dozen lures dangled from my torn rain jacket. I limped from a badly twisted ankle.

By now, Jim was running over. As he retold the story later, he saw me standing on top of the jetty then heard a booming explosion and saw whitewater thrown forty feet in the air. When the water settled down, I had vanished.

By the time I made it back to the dock, pain was starting to report in from various parts of my anatomy. Miraculously, I had broken no bones, though I had a deep gash in my shin and a slightly dislocated jaw. We all agreed it could have been much worse.

Aaron wound up winning that year with a measly two-pound dogfish he caught from the safety of the dock while I was getting twenty-one stitches at the Block Island Medical Center. Clearly I would have taken home the Bunker Cup with a fine striper if only I had made it to the end of that jetty. But it would have been awarded posthumously.

Part IV

CATCHING THE CREEPS

Loon Lake Northern Pike

Creepy Uranium Man

One summer while traveling around Wyoming, I spent the better part of a week fishing a pond known locally as the "Uranium Pit"—aptly named because it was once a uranium strip mine. But it also contained enormous trout, nourished by spring-fed, alkaline waters stuffed with freshwater shrimp that would cling to your waders by the hundreds. Even the water itself had a faint waft of shellfish. No, the trout did not glow.

To get to the pond, you turned off a lonely state highway at a tiny, atrophied mining town. Then you bounced along a dirt road for miles. It was desolate out there; just rolling sagebrush-covered hills studded with a few rocky outcroppings. Here and there, faded signs pointed to overgrown roads that led to long-closed mines. One read "JACKPOT MINE" painted in dripping red letters that looked like it was done by either a kindergartener or Count Dracula. Finally, you drove over one last rise and down a steep hill to a twenty-acre pond carved unnaturally out of a hillside.

The best fishing took place at night, long after the few locals fishing there would reel up their worm rigs, pile into their pickups, and presumably head for the one bar left in the mining-town-gone-bust. The sun would go down and suddenly the weedy shallows would bulge with cruising trout that would graze on shrimp like fat, happy heifers in a hay field.

That first night the fish started out fussy, ignoring every shrimp pattern I threw. Then as I waded out of the pond to change spots, I spotted a small leech clinging to my waders among many dozens of shrimp. So I switched to a marabou leech and began stripping and jerking it through gaps in the weedbeds. Eureka. Trout rushed it like hungry bonefish. At first, the fish I landed were smallish—foot-long brookies and rainbows, but fat as footballs. Then as it grew darker, much larger trout ventured into the shallows. These fish, when hooked, would thrash wildly and dump yards of line off my reel. Two of my personal-best trout—a twenty-inch blimp of a brookie that weighed close to four pounds and a twenty-six-inch, six-pound rainbow that jumped like a berserk tarpon, both came from the Uranium Pit.

On my last trip there, I arrived under a rising full moon. Earlier, I had loaded up on leech patterns at a local fly shop and guzzled a sixteen-ounce coffee prior to what I hoped would be a long, pleasant night of trophy trout fishing.

I parked at the water's edge adjacent to a steep bank that marked the pond's best weedbeds. The locals were long gone. I quietly waded into the water, stopped, and realized that I was hearing something I had not experienced before—a total, utter lack of sound: no crickets; no owls; no airplanes overhead; no cars in the distance. I relished that I was completely alone in the sagebrush of Wyoming.

Or was I? I glanced over my shoulder ready to cast, and there he was—someone standing behind me on the shoreline about fifty feet away. I stopped and stared. Whoever he was remained motionless and just stood there in front of the steep bank. Any romance of fishing the quiet, lonely plains of Wyoming was suddenly gone, replaced by a strong uneasy feeling. I decided to wade down the pond. When I did, the would-be stalker walked silently along the shore with me. So I stopped, and he did, too.

By now I was just plain scared and trying to think of a way out of this. I reached into my pocket and clutched my Leatherman tool contemplating whether I could use it as a weapon—maybe I could stab him with the leather punch. Meanwhile, the silent, creepy, stalker just stood there apparently waiting for me to make my move. And when I moved down the pond some more, he walked silently with me.

Now I was desperate and began sloshing clumsily back to the truck, and creepy uranium man was heading there, too. I had no chance; the mutant would catch me easily, and they would find me the next day floating in the pond covered in shrimp with my eyeballs sucked out by leeches.

And then, just as I reached my truck where the steep bank ended, the uranium freak vanished. I stopped and looked around. He was gone. Then I noticed how the strong glow of the rising full moon had illuminated the bank and hillside behind it. I thought for a second, then craned my head toward the moonlit bank and watched the mutant's own disfigured head appear. When I leaned back toward the truck, he vanished again.

Creepy Uranium Man was my shadow.

I think.

Night of the Living Hellgrammites

One of the best ways to learn the inner workings of a trout stream is to get a small minnow seine or a decent-sized aquarium net and wade into a shallow riffle. Have someone hold the net immediately down-current—just make sure it's touching the stream bottom. Then start turning over rocks. Do this for a minute or two then lift the net from the water. If the stream is moderately healthy, you will now be staring at a seething mass of trout food.

Start picking through your haul to see exactly what's in there. Assuming you've placed your net in a bona fide coldwater trout stream, it's a safe bet that mayfly nymphs of various sizes will be crawling about. Some are mere fractions of an inch—maybe blue-winged olives or this year's first molt of Hendricksons. Others are much larger, mottled, and flat—probably March Brown or Cahill nymphs, a sure sign that the stream runs cold and clean year-round. Maroon-colored *Isonychia* nymphs will flip around like miniature shrimp. Place them in a bucket with some water and they zoom about like little submarines.

Clusters of small twigs and pebbles seemingly glued together may turn out to be the homes of stick caddis—look for a hole in one end and you might see a small head staring back at you. Put them in the bucket with the *Isonychias* and they will eventually start lumbering along the bottom carrying their home on their backs. What might look like craggy, waterlogged beetles stuck in the net are actually dragonfly nymphs—ugly creatures that belie one of nature's most graceful fliers.

Virtually all of these critters are harmless. Even a two-inch stonefly nymph with an abdomen striped like a hornet will merely wander over your fingers fruitlessly looking for a rock to hide under.

But if you find something much larger in the net that looks like it belongs not on a trout stream but in a science fiction movie—the one where the creature-from-another-planet crawls into your ear and eats your brain—you have caught a hellgrammite. They can be as long as your index finger with a meaty, muscular body flanked on each side with wriggling centipede legs. The head is topped with a pair of hooked pinchers that can and will draw blood. They eventually hatch into massive dobsonflies—747-sized bugs that can easily be mistaken for bats when they buzz around trout streams at dusk.

Even their name—HELL-GRAM-MITE—sounds scary, almost like some sort of medieval torture device: "Never mind the iron maiden; place the knave in the hellgrammite."

And if regional nicknames are any indication of temperament, draw your own conclusion from hellgrammites, which, according to a favorite fishing book from the 1930s, are also called corruption bugs, crock hell devils, alligators, dragons, snake doctors, hell divers, flip devils, and water grampus. The last nickname is a bastardization of krampus, a European mythological monster known to haunt misbehaving children.

Except this is no mythological monster; these things are real. And as a group of us learned one late spring night on our annual camping trip on the Upper Delaware River, they will attack.

It was well after dinner as ten of us sat contently around a campfire. Our comfortable—but manly—camp is worth describing: Five canoes rested on the grassy shoreline a few yards from the fire-ring. Off to one side, a camp table served as a combination bar/buffet. Behind that, a row of coolers rested. Periodically one would creak open and you would hear the rush of ice cubes, then the clink of a found bottle of beer followed by the rasp of an opened bottle cap. Fly rods stood upright against trees while waders and fishing vests hung from lower branches. An old oil lantern hung from a hook. It gave the outer camp a warm, golden glow, while the licking campfire bathed the men sitting around it in pleasant yellows and oranges. The smells of woodsmoke and cigars mingled with the aroma of spicy river weeds and dewy grasses. The big Delaware gently purled away. Pleasant chatter, punctuated with laughter, permeated the air. Everyone was happy.

Paul had the misfortune of being closest to the river.

He suddenly shuddered and brushed something off his calf. Headlamps and flashlights immediately trained on a black, creeping mass of legs and jaws on the ground. It was a hellgrammite—a big one. This was followed by a barrage of "What-the . . . ?" and "I-thought-they-only-lived-in-the-river."

I grabbed the corruption bug by the collar just behind the head—the one spot where it can't bite you. Nevertheless, it did its best to pry itself loose so it could sever my hand below the wrist in a single chomp. But before it could do any damage, I tossed it back into the river, where it landed with a surprisingly substantial splash.

We settled back down and naturally began a discussion of hellgrammite natural history, of which it turned out we knew little. We concluded that this one was an anomaly—drawn out of its usual watery habitat by some sort of migratory instinct gone wrong.

Then Paul shuddered again, this time swatting his neck. Something large fell at his feet and again headlamps and flashlights shined down. A crock hell devil—maybe even a little bigger than the first—glared back.

Once more, I volunteered to do the wrestling, but this time I made sure to toss the invader considerably downstream where the current would hopefully carry it far away.

By now some of us had moved a few feet away from Paul, who was constantly scanning the ground around him. Then it was Josh's turn to shudder, this time kicking something off his foot. A snake doctor gnashed jaws that could double as a bottle opener.

By now it was clear that hellgrammites had penetrated the perimeter. All of us began scanning the ground with lights and to our collective horror, dozens of flip devils were converging from the grass all heading to one point—the fire pit.

Whether it was the light or the heat that drew them in, it didn't matter; this was a direct assault on the very soul of our previously described cozy, manly camp. And for the next several minutes, ten rugged outdoorsmen squealed like frightened children being chased by the real-life krampus.

Then one of us had an idea: Let's use them for bait. Though our collective knowledge of hellgrammite natural history was scant, their effectiveness as live bait was legendary. Everything from smallmouth bass, to walleye, to trout relished them.

An empty container was produced and we began filling it with hell divers. In half an hour we had several dozen. We put the container in one of the coolers to use the next day, placing a large rock on the cooler lid just in case.

That night, sleep was difficult—particularly for those who decided to sack-out under the stars or under flimsy tarps. The next morning several of the group reported hearing the clicking of grampus legs walking by their heads. One found a flip devil deep in his sleeping bag—and surprisingly lived.

But now, the hunter would become the hunted—or something like that. The fishermen in the group divvied up the snake doctors into smaller containers. Buck headed upstream where he would drift them in the slackwater for trophy smallmouth. Jim launched a canoe to bottom-fish the channel for giant walleye. I took mine to a swift run downstream where I would feed them to big, German brown trout as long as your arm.

By mid-afternoon we returned to camp to show off our amazing catches. Bob had released a ten-inch smallmouth three hours ago; Jim and I had caught nothing. We stared blankly at each other. We eventually dumped the rest of the corruption bugs back into the river, where we imagined they would crawl away to plot their next dastardly move.

Later, I would learn from a guidebook the more likely scenario: The hellgrammites would simply crawl back ashore and hatch—their original intent before ten goofballs got in their way.

Casting Past the Graveyard

Anglers sometimes speak reverently about a place "where fish die of old age." This is usually a euphemism for some heavily posted or otherwise inaccessible private stream or pond they can only gaze at from afar. It's purportedly full of big, dumb, wild, hard-fighting fish that never see a lure or bait. And the landowner doesn't even fish.

But in reality, few places exist where fish live leisurely, unmolested lives. Even with no human fishing pressure, there is a long list of opportunistic predators—from herons to grizzlies—ready to take advantage of anything long-in-the-fin that shows the slightest sign of slowing down.

Yet southern Chile may indeed be this mythical place—at least for brown trout living in certain inland lakes. Browns are not native here; they were stocked a hundred years ago by the British. And bears, otters, and osprey don't range this far south. The only non-human predators of trout in this off-kilter landscape are other brown trout.

Rodrigo called it the Trophy Pond—a thirty-acre prairie lake impounded by beavers, another introduced species. The trout had colonized shortly after the beavers set up shop. When I asked him the last time it had been fished, he thought about it for a minute then told me seven or eight years ago. It sounded like something out of a dream.

To get there, we drove miles down a rutted dirt road through an old sheep ranch, and then hiked through a wetland of colorful spongy mosses called *turba*. When we reached the pond's shoreline, I began wading toward a channel where Rodrigo remembered big browns sometimes cruised. With each step, clouds of freshwater shrimp scuttled away from me. I tied a small olive scud pattern to my leader and began casting it.

A few casts later I felt a slight weight, like I had picked up a small piece of weed on the end of my line. I lifted, and the rod bent but did nothing else. Yes, definitely weeds.

Folk Art Brown Trout

No, wait—now I felt slight resistance pulling back. No head shakes; just an opposing force that wanted to go the other way but could barely do so.

Then I saw something roll on the surface and could make out a yellowish fin. I had hooked a trout—a large one. But it did little more than wallow. I began reeling in the hooked fish and could now feel the slow-motion pumps of its tail as it tried to swim away but instead was being slowly towed backwards.

Now I could see it: a very long brown trout—well over two feet. But it was all head; its body had withered over time into an afterthought. In its prime, it would have gone six pounds, maybe seven. Now it weighed barely two. I had caught a geriatric trout that had gummed my scud pattern.

I showed the ancient trout to Rodrigo, who was not surprised.

"We call them snakes," he told me. "We sometimes catch them. It probably won't make it through the next winter."

I let it go and watched it swim feebly away. I had just caught the proverbial fish that would die of old age and wished I hadn't.

Be careful what you cast for.

The Brotherhood of the Folded Pants

I have no problem telling anyone exactly where this spot is. Drive north on the New York State Thruway and you can see it: an old steel railroad bridge crossing the Ramapo River. There's a very trouty-looking hole beneath the bridge that's probably worth checking out.

I say probably because I never actually fished it myself. But I once came very close.

The Ramapo is one of those marginal trout rivers overshadowed by the more glamorous waters of the nearby Catskill Mountains. It has the reputation of being a "local's stream," heavily fished by mostly bait guys in the early season, then becoming bathtub warm by the end of June.

The river runs through the valley of the somber Ramapo Mountains—part of the Appalachian chain. According to local lore, the Ramapos are home to the "Jackson Whites," an enclave of mountain people who trace their ancestors back to escaped slaves, Native Americans, and Hessian soldiers that deserted during the Revolutionary War. Some swear the Jackson Whites are still there eking out a subsistence living just twenty-five miles from New York City. They warn you not to drive up certain dirt roads leading into the mountains.

All of these factors gives the Ramapo a patina of funk that reaffirms why I had passed the trout hole a hundred times without stopping.

But the day I nearly fished it was different. It was shortly after opening day of trout season and most of the upstate rivers remained too cold to fish. The more southerly and heavily stocked Ramapo seemed like a safe early-season bet, so I decided to try it.

I began in a popular stretch that flows near the town of Sloatsburg. It was a warm spring day, and the river ran unusually low and clear. Several spin fishermen wandered around impatiently casting from spot to spot. No one was catching anything. I nymphed through a few promising runs but came up empty, too.

So I got in my car and headed downstream following the river along the old state highway that parallels much of it. The Thruway, too, was never more than a few hundred yards away.

And that's when I spotted the old railroad bridge. I recognized it immediately and remembered the deep hole beneath it. I parked at a turnout just downstream of it and pulled on my waders to the roar of the highway. No fishermen were in sight, but there was a surprisingly well-worn path leading directly to the pool. Perhaps I had stumbled onto someone's secret spot.

The pool itself looked deep and mysterious. Riprap boulders lined the banks and spilled below the surface before fading into bottomless green water. It screamed big trout.

My plan was to fish the pool starting at the head. I would use a large Woolly Bugger—a good searching pattern for aggressive, early-season fish. If that didn't produce, I would bounce the bottom with a heavy stonefly nymph.

To reach the top of the pool, I had to hunch down and make my way under the bridge. It was dark and noticeably cooler there. The sound of water lapping against the riprap boulders echoed off steel beams making weird metallic sounds.

That's when I saw the pants.

They looked clean and were neatly folded and laid carefully on top of a concrete support. They were green. They didn't look like they had been there a long time—maybe a few days or maybe a few minutes.

I stopped and glanced around. Was the owner about to come down the path with the rest of his laundry? But no one did.

Now what should I do? A beautiful, mysterious unfished trout pool beckoned. But whose damned pants were those? A troll's? Did I blunder into someone's living room? Were the pants from a Hessian uniform? Was it a trap set by the Jackson Whites?

I stood there hunched underneath the bridge thinking. One time on the lower Brodheads Creek I passed a homeless person heading for his primitive shelter on the banks of the remote trout pool I had planned on eventually fishing. Then there was the day on the lonely Pohopoco Creek in Pennsylvania coal country when I saw a man dressed in little more than rags staring at me before ducking into what looked like an abandoned mine.

Those happened years ago. Now, I was clearly much more seasoned, with graying jowls in my beard and temples. Maybe this time I would just fish the pool like a man and to hell with whoever owned the neatly folded green pants tucked under the damp, dark bridge.

Not a chance. Less than two minutes later, when I made it back to the car, I didn't even take off my waders. I just climbed in and drove directly to the on-ramp of the Thruway and headed home. My fly rod lay in the backseat, the uncast streamer tangled in my leader.

This was a couple of seasons ago. The pool is still there, tantalizing and swirling away beneath the old steel railroad bridge. I've passed it dozens more times and have yet to see anyone fishing it. There might be some really nice trout in there. If you go, let me know how you do.

Pleasure Boating

Scratch beneath the tough veneer of a self-described hardcore surf fisherman—the kind that waxes on about rugged individualism and how standing toe-to-toe in a booming nor'easter is the only way to catch a fish—and you will find a puker; someone who vomits whenever they set foot on a boat.

I know because I'm one of them.

I shouldn't say I puke every time. In fact it's been a while since I've "chummed"—as it's known in the charming vernacular of hardy boat fishermen. But it's happened enough to make me generally leery about boats. Invite me surf fishing and I will be there regardless of time, place, or conditions; invite me on your boat and I need to check vital signs first—namely the wind speed and wave forecast, followed by my blood pressure and heart rate.

Which is why I'm still scratching my head about the scariest forty-five minutes I had ever spent on a boat—or anywhere for that matter. You see, we knew the weather would be bad—that the winds would increase to near gale force from the southwest and the seas would build. But the four of us went anyway.

I partially blame it on Block Island, that fishy pile of glacial till twelve miles south of the Rhode Island mainland. Avoiding sloppy seas by knowing where to tuck out of the wind is a high art there.

The plan was to fish in the morning on my friend Paul's boat for a few hours in the eastern lee of the island, then retreat back to the harbor on the western side before the weather deteriorated. We knew the last part of the run home might be snotty as we would be heading directly into the wind. But it was only two miles; how bad could it be?

The day started out fine. Tim, Aaron, and I piled into Paul's twenty-four-foot cuddy cabin and headed out of the inlet on an ebbing tide,

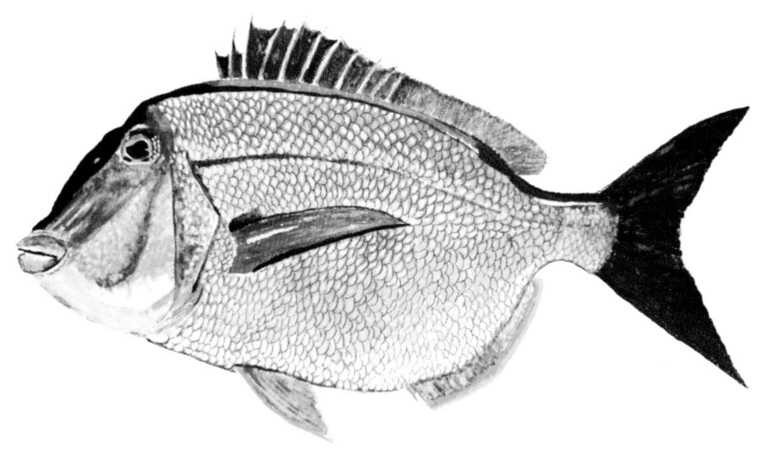

Block Island Scup

then turned north into a relatively gentle following sea. Then we rounded the famous north rip and headed south staying behind a series of cliffs. Along the way, we slow trolled tube-and-worm rigs looking for stripers. This is finesse fishing; you creep the rig past submerged rockpiles trying to coax out a big bass. If it works, strikes are violent.

But we found nothing. So we motored farther south to jig for delicious black sea bass off a patch of rocky bottom. Our jigs were intercepted almost immediately, but it turned out we were pillaging a sea bass nursery with no keepers landed and most fish well under eight inches.

That's when we began to notice that our drifts were getting faster as a freshening breeze started pushing us seaward. We moved in closer to the cliffs and it didn't seem so bad, so we naturally stayed longer. We ate lunch in the warm beguiling sun, not noticing that the wind speed continued to increase.

After lunch we jigged some more for sea bass, but it became clear that the larger fish were just not hitting. Maybe the dropping barometer from the approaching southerly gale had turned them off.

By mid-afternoon, we decided to head back to the harbor. Our intention was again to stay tucked into the leeward eastern shore, then round the north rip and head south into the slop for the rest of the way.

The one factor none of us had thought much about was the tide. It had started to come in, and now was surging headlong into the strengthening sou'wester. This causes the seas to build . . . and build.

As we motored north and the cliffs gave way to open beach, we could now begin to feel the full brunt of the wind. It was honking, quartering in over our shoulder at a good twenty-five knots. Meanwhile, the tide was hurrying us into the shoal water of the north rip where the currents and wind were about to go to war.

Paul, God bless the man, finally broached the idea of turning around. Maybe we could hole up in Old Harbor on the protected eastern side of the island for the day, he suggested. We would have to take a taxi back to the house and then retrieve the boat when the weather subsided. There was a brief discussion about the pluses and minuses of this, namely its inconvenience and extra cost for the docking fee along with cab fare. All the while we were slowly heading closer to the rip.

But in the two minutes it took us to all agree that turning around might be the prudent thing to do, we could no longer do so. The tide had pushed us head-on into waiting waves that heaved upward from the gale and rammed against the incoming tide. We hit the first wave and the boat crashed violently downward then pitched upward, sending a wall of whitewater over the cabin. We were ambushed by an army of eight-to-twelve-footers—daring us to turn around so they could hit us broadside and flip the boat. Paul told us as calmly as he could to put on our life jackets. Mine had been on for the last ten minutes.

And then, to quote Scout from *To Kill a Mockingbird*, thus began our longest journey together.

It was chaos. The boat charged up the faces of wind-streaked twelve-footers only to plummet into the valleys behind them. Sometimes the bow would bury into the trough of a particularly big wave then resurface throwing a hundred gallons of seawater over the cuddy and into the cockpit where it would rain down on us and sound like thousands of ball bearings bouncing off the deck.

That's when Tim called out: "Taking on some water back here."

I was literally too scared to turn around. Plus it would interfere with my new self-appointed job, which was to spastically call out "BIG WAVE!" as if Paul couldn't see another twelve-foot battering ram bearing down and about to knock us silly.

Paul didn't turn around either; he couldn't. He replied in a tone that may have been more to reassure himself: "Those self bailers should be working. It'll drain." We pounded into another wave and the bailers thankfully did their job.

Meanwhile Aaron, who stood behind me, remained silent. Either that or he had been swept into the Atlantic long ago. Again, I wasn't turning around to look. Later, I would find out he was madly texting his fiancée telling her how much he was going to miss her.

As we rounded the point and headed southward, I watched the shoreline move past at a painfully slow rate. We were just barely making headway.

The pounding and beating went on. For the most part none of us spoke. If Paul climbed up and over a particularly big wave then coasted down and turned without burying the bow—no easy feat—Tim would call out, "Nice job."

All the while, the sea—that cold, uncaring witch—continued to batter us. Paul's boat groaned and heaved. The harbor breakwater, less than two miles from us, seemed as far away as Pluto. But slowly—excruciatingly slowly, painfully slowly—it grew closer. And closer still. Pluto became Neptune, then Jupiter, then Mars.

Then, after what felt like the length of the Shackleton Expedition, but was no more than forty-five minutes, we finally passed the breakwater into the safety of the harbor.

In the calm waters of Block Island's Great Salt Pond, we all looked at each other shaking our heads over what had been a terrifying experience. Paul admonished himself for not turning around when he could. We just thanked him for getting us back to the dock alive.

We returned to the house and all of us went straight to the cooler for stiff drinks. Then while gazing at the cruel, storm-streaked Atlantic, I realized I had made a medical discovery important to all surf fishermen: you don't puke if you're scared shitless.

Part V

BLOOD KNOTS

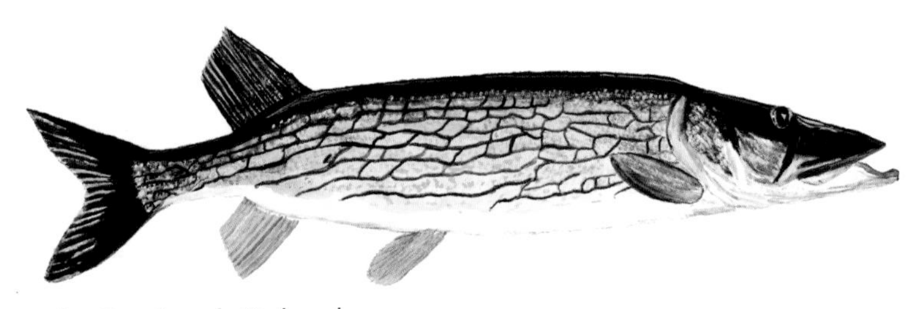

Lake Oquittunk Pickerel

Jacked Up

Any serious angler will tell you that mixing serious fishing with family "quality time" generally assures mutual disappointment for all involved. This is why things like fishing clubs, "just-the-guys" annual fishing trips, and boats without bathrooms were invented.

On the other end of the spectrum, it may be why cruise ships to Bermuda were invented. The week-long cruise was a gift from my parents to my two brothers and me and our respective wives and kids: eleven of us in all. Relaxing by the pool. Gambling. Overeating. Shuffleboard. Fishing was not on the itinerary.

But how could I not at least pack a three-piece spinning rod just in case there was some convenient place to cast nearby? Heck, I probably wouldn't even use it. And just to be safe, I should also bring an eight-weight fly rod. Plus a box of bonefish flies. And jigs—lots of jigs—I'll definitely need those. Better add a few big plugs. Oh and my favorite poppers—can't leave those behind. Spare reel? Check. Extra spools? Check. By the time I was done, a large backpack full of fishing gear bulged at the seams.

At the cruise ship check-in, while others tried to sneak on their own personal supply of booze to avoid paying inflated bar fees, I was trying to waddle past my family lugging a small tackle shop.

Once on board, my wife and I quickly learned that there are two kinds of people in this world: cruise folk and normal humans. Cruise folk like to eat and drink in large quantities, get massages and facials, and lay in the sun like basking iguanas (some had the skin to match). I contemplated trolling off the stern but thought better of it.

When we arrived in Bermuda, I did try fishing briefly off one of the cruise ship's upper decks one night, but found the eight-story drop to the water not very conducive to landing anything.

The next day, my parents organized an outing to an out-of-the-way beach they heard about on the far side of the island. To get there, you had to take a bus to a ferry that took you to the city of Hamilton, followed by another bus and a short walk to the beach. I of course packed a rod and a small tackle box—just in case.

It turned out to be a lovely spot—a lightly traveled beach with swaying palms and pinkish sand. We swam, played with our kids, and relaxed for hours. Wives read contently and grandparents sat in the sun. My pack rod and tackle remained forgotten in the backpack.

Then, tired, sandy, and a little sunburned, we all took the bus back to Hamilton chatting all the way about what a fine time we all had. At the dock, we learned that the ferry would arrive in half an hour, so some of my family began to wander into a few nearby shops. I stayed behind holding my three-year-old son who was sleeping over one shoulder; over the other hung the backpack with the pack rod and tackle. Before my wife left, she asked me if I was OK, and I assured her everything was under control.

And it was. I gazed over picturesque Hamilton Harbor shimmering in the afternoon light and filled my lungs with semi-tropical salt air. My son looked like a sleeping angel.

Then the jacks showed up.

I heard them first—a sound reminiscent of a thousand people clapping—except louder. I looked around confused, trying to locate the origin of this strange yet familiar noise. And then I saw it. Half an acre of Henderson Harbor was being torn to shreds twenty yards off the dock.

Hundreds of fifteen-to-twenty-pound jack crevalle—among the hardest fighting fish in the world—were absolutely destroying schools of baitfish. Never had I seen such a violent feeding frenzy. All I needed to do was string up my rod and make a cast with any jig, plug, or spoon in my tackle box. A big jack would immediately crush the lure then give one of the all-time great fights of my angling career.

I excitedly turned to hand off my son to one of my family members. All of them were suddenly gone. Frantically, I walked up and down the dock looking for them, but they were nowhere to be seen. All the while the jacks strafed and marauded.

I quickly weighed my options. Maybe I could prop up my sleeping son against a pier piling and have at the jacks; but I quickly dismissed the thought. Then I considered reaching behind me and somehow grabbing the rod from the backpack, assembling and rigging it, casting a lure, and then hooking and landing a twenty-pound jack—all with one hand and while holding a sleeping three-year-old. But since I couldn't even complete step one and reach the rod, I decided to scrap the rest of the plan.

This was getting serious. Maybe my family had played a joke on me and somehow snuck back to the ship to hit the blackjack table. Meanwhile my son continued to sleep away nuzzled against my shoulder and starting to feel like a giant swollen tick.

The jacks pinned the baitfish against the bulkhead a few feet from me and wreaked utter havoc. If I had a net I could have scooped one up. The blitzing went on for a few more minutes, with me frantically searching for my family who had clearly abandoned me. Eventually the school dispersed. I stood, shoulders slumped as much from watching the fish leave as from the five-hundred-pound weight I now felt like I was carrying.

Then everyone came back. Some had been to a nearby restroom; others had poked around a few stores. Some had ice cream cones. How nice.

My wife could see I was sweating, though it wasn't from the heat.

"Do you want me to hold him for a little while?" she asked just as the ferry arrived.

"No, I got him," I said, knowing he was the only thing that could distract me from an unscratched itch that felt like a thousand ants swarming over my body.

Fishing with Girls

This is touchy stuff, I know. And when I say "girls," I use it in the vernacular referring to any non-fishing husband, wife, boyfriend, or girlfriend. So don't judge. Yet.

Here's another disclaimer: My wife fishes and is the only person I know who caught a twenty-five-pound striped bass and a five-pound striped bass on a single cast (the big fish took a plug; the smaller one grabbed a teaser). So there.

But there were others before her, I freely admit.

In theory, it all sounds perfectly honorable: You want to share your passion with that special someone—blah, blah, blah. But in practice, maybe you're just looking for an excuse to sneak in a few more casts.

In college, I tried hard to get a particular girlfriend to partake in the pleasures of the angle by taking her to a local pond where she caught some bass. She seemed to enjoy it, sort of. Though the word "tolerate" also comes to mind.

Then it dawned on me why her reaction may have been only lukewarm: She hadn't yet experienced what I believed at the time to be life's most fulfilling moment, namely catching a bluefish on a topwater plug in the surf. I know what you are thinking, but I was young and immature. I've since grown up and now know that the most fulfilling experience is, of course, raising a big wild brown trout on a dry fly. Or maybe a salmon. Hmmmm.

Back to the bluefish . . . I mean my girlfriend. I waited for the perfect moment: mid-May, when a man's fancy turns to love and bluefish gorge on bunker 'til they puke. I chose the point of Sandy Hook in northern New Jersey with its panoramic views of New York City. Of course I had to pre-fish the spot—several times—until I had found the best lures and correct tides. Got some nice fish along the way, too, including a fat ten-pounder that jumped six times. Damn nice fish.

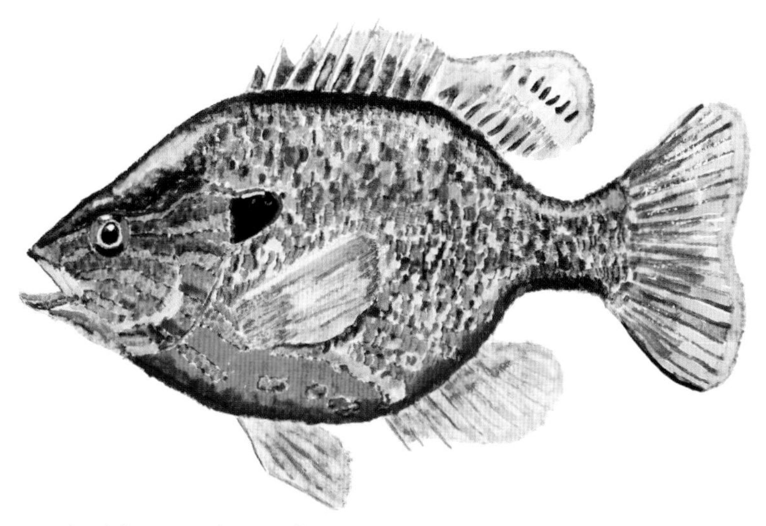

Psychedelic Pumpkinseed

The fateful day came. We hiked to the point of the Hook where its famous rip was pouring its heart out with the ebbing tide. I could almost picture the schools of baitfish beneath the surface being chopped into little bits by marauding blues. It was beyond romantic.

I handed her a surf rod perfectly rigged with my second-favorite popper. My favorite, of course, was strung to my own trusty nine-footer. No one but me gets to cast that one you understand. Did I tell you about the ten-pound bluefish I caught with it?

I instructed her how to make a long cast and she did. Then I showed her how to retrieve the lure, making it look like a wounded baitfish.

On maybe the third cast, the water humped up behind my plug. I kept retrieving. There was a slash and a miss by an unseen fish. I cranked some more, prompting a violent strike that bent the rod deeply. Before I could proclaim "fish on!", I turned just in time to see my girlfriend hook her own bluefish.

Paradise found.

We landed them both, sleek four-pounders, and I instructed her to cast again, that there were plenty more out there. She did and I did, and we hooked up once more and again landed another brace of beautiful bluefish. Our next casts met with the same result. The love theme from *Romeo and Juliet* played in my head.

I pictured our sixth, seventh, and eighth casts together and into blissful eternity.

Then she turned to me and spoke these fateful words:

"How much longer are we going to be here? I'm getting kind of hungry."

The needle scratched off the record player.

I gently protested—in a romantic way, of course—telling her that baby, this fine bluefishing should satisfy you more than any mere meal.

She didn't see it my way.

So with a rip full of hungry bluefish, I trudged off the beach with her. She got her meal; I was starved of mine.

Another time and another girlfriend later, I had just returned from a solo fishing trip for shad. I had caught none, but saw a remarkable sight: In an upwelling of whitewater below a submerged boulder, a very large shad briefly appeared in a plume of bubbles. The fish shined purple and silver in dappled light before fading into the depths. It was a moving sight, and I called the woman I had just started dating and gushed about this amazing visual image. Then I suggested that she should come fishing with me, so she, too, could experience something so beautiful.

There was a brief silence on the other end of the line.

Then she said: "You want me to go *fishing*?" as if I had asked her to visit North Korea. At that moment I knew that this too, shall pass. And it did.

Then there was the time on opening day of trout season a few years later. I had access to a private and heavily stocked stream and took yet another girlfriend there to catch our limits.

Somehow on the walk to the stream we got into an argument. I don't remember exactly what it was about—maybe something about me running half a mile ahead of her to reach the spot. Whatever it was, we wound up in a verbal donnybrook followed by her crying on the riverbank.

With a stream full of heavy stockers just begging for a well-cast gob of worms, I found myself sitting next to her apologizing for whatever it was I may have done. Long minutes later, we made up. We hugged with me peering over her shoulder wild-eyed looking for rising trout.

I promised her it would not happen again. And I kept that promise. By the next opening day, she had dumped me and I caught my limit.

My friend Jim may have figured it all out when it comes to achieving that delicate balance between fishing and girls. He brought his non-fishing wife on his boat on the lower Delaware River one April afternoon. It was a warm, pleasant day and she was just happy to be there reading a book and enjoying the sunshine.

Then one of Jim's rod's goes off and a huge shad—the shad of a lifetime—leaps.

Jim grabs the rod and begins fighting the fish, but it's too powerful. Yards and yards of line are pouring off the reel, so Jim clambers to the bow of the boat and manages to pull the anchor with one hand while continuing to fight the fish with the other.

It was a dramatic and heroic battle—both for the fish and angler. There were jarring head shakes, more spectacular leaps, and runs that nearly melted the bearings of Jim's reel. Almost a mile of river had drifted past during the fight. Eventually, Jim got the upper hand. He eased the shad closer then grabbed his landing net. The shad rolled on its side, exhausted.

He brought the fish closer, then closer still, then finally over the lip of the net. Then with a quick lift, the shad was his. It was truly enormous—a massive female. Later, at a tackle shop that was holding a fishing contest, it would weigh more than seven pounds.

Nearly half an hour had passed since the rod first went off. He turned to show his wife his trophy—the largest and greatest shad he had ever caught.

She was curled up in the bottom of the boat sound asleep.

Jim wound up winning first prize, and his wife got in a refreshing nap—probably the best possible outcome for all involved.

The Mother's Day Trout

The new four-weight was burning a hole in its aluminum tube, just dying to be cast. It was my first-ever split bamboo fly rod. Seven-and-a-half-feet long. Honey colored and with a burled reel seat. It had the kind of slow, old-school action that harkened back to wicker creels and field-dressed brook trout lying on beds of ferns. I had just bought it from a local bamboo rod collector/maker who occasionally built one for himself. And now on a warm May afternoon with a high overcast and no wind, I knew the perfect trout pool where this rod simply must be fished.

One problem: It happened to be Mother's Day.

Robert Traver wrote in "Testament of a Fisherman" that he loved to fish "because of all of the television commercials, cocktail parties and assorted social posturing I thus escape." Regarding his last point, either he had the world's most understanding family or I'm calling bullshit.

Growing up in a non-fishing family, missing a social gathering because the tide is perfect, or the shad are running really thick, or the Hendricksons are hatching was not acceptable. I would make impassioned pleas about the incredible importance and rarity of these phenomena. In return, I would get blank stares.

Depending on the type of gathering, I'd sometimes go fishing anyway. A casual get-together could get trumped by a good bite of stripers. My philosophy became fish first, apologize later. Few understood, except my adoring grandmother who used to exclaim: "At least he's not in some tavern!"

But certain events and holidays were command performances. My brother's wedding, for example, though I did stow a rod for after the ceremony.

So on this Mother's Day, after a leisurely lunch, opening of gifts, and a few hours of pleasant conversation, I felt I had sufficiently celebrated the day and honored my mom. So I stood up, stretched, and made overtones that I would be leaving soon.

"Why? Where are you going??" both brothers demanded.

They already knew the answer.

I could have gone on about the new bamboo fly rod, how the barometer was just starting to fall, that stream levels were perfect, and how I thought mayflies might just be starting to hatch on the Brodheads Creek. Instead I just said: "I'm going fishing."

My brothers gave me that bemused look I had come to know so well. However, my parents—especially my mother—were gracious. After all, they had been dealing with my affliction for many years. They both wished me good luck.

An hour later, I skidded into the turnout along the stream in a spray of gravel. I yanked on waders, slung on my vest, and was assembling the bamboo rod mid-gait as I headed to the pool.

When I got there, I was disappointed to see only the dimples of a few fallfish at the tailout. Some caddis and the last of the Hendricksons flitted about, but no real hatch to speak of. I knotted on a Dun Variant—a sort of generic gray fly that imitates several types of bugs. I figured I would cast to the fallfish just to get a feel for the rod—who cares if it turned out to be a six-incher?

I false-cast and the rod flexed elegantly. Line shot toward the closest rise. The variant touched down then disappeared in a telltale dimple. But when I set the hook, instead of a small chub, a very large brown trout hurled itself from the pool. The four-weight bent into a deep C-shape while the fish took off on a long run and jumped again. I slowly backed out of the pool playing the trout as carefully as I could.

The rod performed just as I had hoped, absorbing the big fish's violent head shakes and bores for the bottom. Several minutes later, I slid the trout into the shallows and quickly laid the new rod next to it for scale. I clicked a picture, twisted the hook free, then watched it swim away.

I still have the faded, grainy photo of a beautiful nineteen-inch butter yellow wild brown trout lying on its side next to an equally beautiful bamboo fly rod. Caught on Mother's Day—a perfect thank you gift from a mother to a fish-obsessed son.

Bowling for Shad

On certain Delaware River tributaries you can still find remnants of American shad runs. Some are predictable and can be targeted each spring for a week or so. Others seem to ebb and flow over the years depending on the right conditions. Sadly, all are mere ghosts of their historic abundance when virtually every East Coast river had a run of shad, herring, striped bass, or sometimes even Atlantic salmon. Still, it's a nice surprise when you cast, and instead of a foot-long trout or smallmouth, a leaping six-pound, sea-run fish is on the other end.

I once found such a place. It was a marginal trout stream not known for much of anything. I'm not sure how many shad ran up it, but it seemed like less than one hundred. You would spot pods of two or three here and there in a few of the deeper runs as they made their way upstream. Most of the fish would eventually wind up in a single large pool where the river dead-ended below a tall dam that blocked further migration. Here, they would swim in larger schools of a dozen or more. And, as I happily learned, they would slam a fast-stripped streamer if you could get a cast off in front of the leading fish.

One day I took my wife, Mimi, there. It was a steep hike down a trail that led to the tailout of the big pool. Then you would slip and slide upriver over algae-slickened boulders and cobble before you reached a rocky outcrop midway up the pool where you could cast.

The pool was deep and mysterious with brown, tannin-stained water. There were rumors of walleye and big trout lurking in its depths. Fifty yards away, the dam itself stood massive and imposing. On higher spring flows, sheets of water would crest over the top, free fall one hundred feet in a curtain of white, then slam off a concrete apron with such force it was hard to hear anything but the roar of water. If you waded too close, billows of mist soaked you.

Delaware River American Shad

We made it to the outcrop and began casting. Soon we saw a few shad speed by—large blue-gray shapes in the brownish flow. A few minutes later, Mimi hooked her first. It fought hard, leaping and running all over the pool, even though the fish was now more than one hundred miles from saltwater. Eventually she released a large buck. A few minutes later, she hooked and landed a second shad followed by a third. The pool seemed full of them.

The fourth fish fought differently, running in short-but-faster bursts. Rapid head shakes thumped the rod. Then we saw a streamlined golden shape surging in the current. The fish turned and we could now see it was a huge brown trout. A few minutes later, it, too, was landed and released—a twenty-three-inch, three-pound monster.

By now Mimi was beaming; clearly this was turning out to be a wonderful day for her—filled with fresh, hard-fighting shad and now one of the largest trout she had ever caught.

I remained fishless. So I decided to wade closer to the dam where I hoped to get off a decent backcast and allow the fast-sinking line to settle deeper into the pool where the shad held. Tentatively I approached the dam, which roared louder and louder the closer I came. When I got about one hundred feet from it, I turned downstream and began casting.

Admittedly, it was not a particularly safe spot. With the roaring water, rushing current, and soaking spray, it felt more like casting for stripers from the end of a jetty during a nor'easter than fly fishing for shad on a trout stream. I worked the fly with fast strips but still hooked nothing.

Suddenly a loud crack went off behind me. I wheeled around, half thinking the dam was starting to go. Instead I saw a large, round, red projectile ricocheting toward me. It landed fifty feet away in a geyser of spray reminiscent of a cannon shot fired from a man-o-war.

I turned back to Mimi who had a puzzled look on her face. Then I waded to where the projectile landed. There, in two feet of water, a bright red bowling ball lay on the bottom of the river. A four-inch divot was missing where it had ricocheted off the dam.

I looked up and could see two heads peering over a guard rail at the top of the dam. A second later they vanished. Then I saw a car speed away.

I reached into the river, grabbed the bowling ball, and brought it to my wife. She stared in disbelief.

We speculated what happened: Mimi thought it was probably some bored teenagers who heaved the ball off the dam but didn't know we were there. When they saw us, they got scared and left. She theorized that they had the ball rolling around in their car for weeks and finally decided to unload it. Then she made another cast.

Clearly, this was a woman whose judgment was temporarily clouded by a fine morning of three shad, a very large trout, and the promise of more. Or this was someone who never had a blunt object thrown at her while fishing.

You see, I had not been so lucky. This was not the first time someone had tried to take me out mid-cast. Years earlier, when I was bass fishing along the shore of an urban lake, I came upon a group of rowdy kids drinking on a hill in the woods behind me. Though I tried to quietly fish around them, they must have seen me, because rocks began to hail down where I was casting. Either they were trying to scare the bass, or they were trying to hit me. I didn't stick around to find out which one. Since then, I still sometimes find myself fishing in certain places with one eye over my shoulder.

So to me, the bowling ball was nothing short of an assassination attempt. When I explained all this to Mimi she remained skeptical. Three shad and a big trout will do that to you.

We continued to debate over the din of rushing water. Whatever the case, I said, I was not waiting around to see if the alleged teenagers were coming back. Rocks and bowling balls will do that to you.

She relented, reluctantly, and we waded out of the pool, hiked up the trail to our car, and left. Sometimes she still brings it up, convinced that she would have landed dozens more shad and possibly an even larger trout that day if only given a chance. Yup, and the spot would now be known as Widow's Pool.

I returned to the pool several times over the following years, but oddly haven't seen a shad there since—or in the entire tributary for that matter. Maybe the bowling ball scared them off, too.

The Gentleman Angler Quiz

See if you can take this quiz, as I recently did, to find out if you are a gentleman angler:

Let's say you are sympathetic to your spouse's desire to occasionally fish. So, once a year, you do the proper thing and take her to a prime trout stream, hand her the rod and her waders, point her to some fine pools, and then in a truly selfless gesture, you take your young son and romp into the woods sans tackle while she fishes.

You allow a full twenty-five minutes of this luxury before you decide to check on her to make sure she's casting properly, and maybe see if there are fish rising—just so you can point her to a good spot, of course. When you find her in the stream happily casting away, you learn to your horror that your wife has not only failed to raise a thing, but she also seems oddly content to be merely fishing.

But you are a gentleman so you insist on putting her in a particular pool that nearly always has a fine trout in-waiting. You hand off your son—temporarily of course—and make a beeline for said trout-lie about a half mile upstream.

And there it is: a foamy plunge pool below a cleft between two boulders. Here, conflicting currents eventually organize into a bubbly foot-wide feeding lane. Many fine browns have come from there, including a hook-jawed eighteen-incher that took a big Adams one day in late June during a hatch of . . . but I digress. This is about being a gentleman.

By the time your spouse catches up now holding your son, you are standing on the bank anxiously waiting for the moment of marital bliss when the husband shall guide his loving and appreciative wife to a fine trout as junior cheers on Mommy and Daddy. You know: husband-of-the-year kind of stuff.

Again you hand off the rod. Then you and the little guy sit on a rock and watch your spouse attempt to fish the pool where a fine trout surely awaits as has previously been noted. Did I forget to mention that this particular lie can only be reached by wading through a fast run studded with algae-slickened rocks? Also: that this particular spouse is a foot shorter than you—meaning that water waist deep to you is chest deep to her, etc.? But this is splitting hairs.

In any case, despite several attempts to cross, your spouse cannot make it. You gently encourage her to try again and again, even fetching her a makeshift wading staff out of a stick, and prodding her back into the river with it. But to no avail. "It's not worth it," she finally announces.

Not worth it?

So here is the question:

Does the gentleman angler:

(A) Tell your spouse it was a good try anyway and suggest some gentler water upstream or downstream that doesn't require such demanding wading, then you take your son back into the woods for some more father-son quality time?

Or

(B) Since your wife wasn't able to wade to the spot, you decide to show her how it can be properly fished by gently but firmly taking the rod from her, while stripping down to your underpants, then wading into the stream in your socks and previously dry sneakers. Then—purely for instructional purposes—you make the cast she couldn't make, and immediately hook a high-jumping fourteen-inch wild brown. When you turn around to proudly show off the fish before you release it, your spouse's expression can best be described as deep concentration (non-anglers may misinterpret this expression as deep anger). Obviously she is making mental notes of how to properly fish the pool when you hand her back the rod again next year.

If you answered B—congratulations! You are a gentleman angler.

If you answered A—you really need to stop telling fish stories.

Part VI

IN FOREIGN WATERS

Cordova Alaska Dolly Varden

Fishing on the Job

Let's say you're stuck at work daydreaming about fishing. All your tackle is home. Would you rather be: (A) sitting in a cubicle; or (B) sitting on the banks of one of the great wild trout rivers in South America? Remember, you are working; either way you cannot make a cast.

I can speak to Choice B. I once found myself in this strange predicament on the Chilean side of Tierra del Fuego in a newly created private wildlife reserve. In many ways, it was a dream trip. A group of scientists, donors, and a reporter were spending the day getting shuttled by helicopter around this Rhode Island–sized wilderness, zooming over glaciers, old growth forests, and grasslands full of guanacos—wild cousins to llamas. My job was to get the reporter what he needed.

We flew over rivers, too; lots of rivers; rivers that were probably full of sea-run brown trout; rivers that had probably never been fished. But this was a work trip, not a fishing trip. There were interviews to set up and schedules to keep. Much as I had been tempted to bring a pack rod, my tackle was at home six thousand miles away.

On our first stop, the helicopter dropped us along the shoreline of a remote bay for a photo-op and to discuss the region's wildlife. This area in particular is known for Andean condors, the world's largest raptor. In fact we could see a few of the 747-sized birds soaring in an updraft over a nearby cliff.

The reporter interviewed one of the scientists while my eyes wandered to a good-sized stream pouring into the bay one hundred yards from us. Above the stream mouth, a rushing riffle tumbled from a mysterious, deep, foamy pool. If ever there was a place for schools of sea-run brown trout to stage before heading upstream, this was it. But all I could do was gawk at it slack-jawed, drool collecting in the corners of my mouth.

Mercifully, the helicopter arrived and it was time to go. We lifted off, and I gazed longingly at the nameless trout stream knowing I would probably never see it again. It disappeared behind the next mountain range.

The helicopter then dropped us off at an abandoned sheep ranch—also within the borders of the new wildlife reserve. Then it left to pick up others from our group. An old homestead sat on a bluff overlooking the headwaters of the Rio Grande—considered the finest sea trout river in the world on Argentina's side of the island. Here in its headwaters in Chile, it ran the width of an average trout stream winding and twisting through a grassy valley that could have been western Montana. With every turn of the river, mysterious undercut banks beckoned.

We waited for the helicopter to return. From there, three of us including the reporter and myself would be heading across the Straits of Magellan to the city of Punta Arenas and eventually back to Santiago. Others were staying behind to continue the tour.

It was a cold, raw afternoon with the wind freshening across the grassland. Sleet started to fall. The reporter was scribbling down notes. So I hunkered down in my rain gear resigned to an uncomfortable wait. That's when I saw it: One of the group pulled a four-piece spinning rod from his backpack.

Immediately, I sidled over and became his best friend, asking about his tackle and his fishing knowledge. It turned out he had never really trout fished and had bought the rod on a whim along with a rudimentary assortment of spoons and spinners.

He made a few casts while I stared lustily. Then, perhaps sensing I wanted it more than he did, he offered me the rod. This was no time for polite modesty so I happily obliged, thanking him as if he had just loaned me one of his kidneys.

I knew I only had minutes before the helicopter returned, so I hurried downstream and began casting the clunky spoon and letting it tumble into one of the deeper undercuts on the far bank.

On the second or third cast, the lure slowed as if hooked into a submerged stick. I raised the rod and the snag churned on the surface in the form of a very large brown trout. The water was cold and the fish sluggish, so it thrashed and bulldogged rather than running or jumping. After a few more surges, I slid it on the bank—a hook-jawed, twenty-four-inch, five-pound trophy. A few quick pictures were taken before I let it go. I was stunned; I had just landed one of the largest trout I had ever hooked in just a few casts. I could only imagine how good the fishing might be if I had more time to explore, or had snuck a fly rod with me along with a box of bright weighted streamers.

Then the helicopter arrived. I handed the rod back to my new best friend who was staying behind. I thanked him over and over before strapping in for the hour flight to Punta Arenas.

I must have still been exhilarated about my catch, because I didn't notice the worsening weather as we approached the Straits of Magellan. Minutes into the twenty-mile water crossing, we hit a snow squall that quickly developed into a full-blown gale. The helicopter began pitching and heaving. A thousand feet below us, icy waters streaked white with foam rolled and crashed.

I gripped the seat white-knuckled and watched the pilot and co-pilot maneuver as best they could. I could hear them through my headphones speaking in Spanish. They seemed composed, but since I didn't understand them, I could only speculate what they might be saying. Were they calmly asking if their life insurance policies were paid up? Maybe they were discussing funeral arrangements?

It was then I remembered the reporter was a fluent Spanish speaker, so I looked over to see how he was reacting to the pilots' dialogue coming through the headphones. He seemed utterly calm even to the point of disinterest. I immediately took solace in his cool temperament while the helicopter continued to labor. If he wasn't scared, neither was I.

And eventually the squall subsided and we could make out the twinkling lights of Punta Arenas. Five minutes later we were at the airport offloading our gear as the helicopter idled. Never had solid ground felt so good.

As we carried our gear inside, I turned to the reporter and thanked him for keeping me from panicking on such a harrowing flight. He seemed puzzled, so I explained how I had watched him during the squall, and how I knew we weren't in trouble because he could understand every word the pilots were saying.

With a bemused look on his face, he said to me: "My headphones were broken; I didn't hear a word."

Then he walked inside.

Thoughts of a watery death, big trout, and unfished rivers swirled around my head as the helicopter took off in the distance. It was a tough day at the office.

The Elephant in the River

I've been chased off rivers by lightning, gale-force winds, clouds of biting insects, a bad cast into a streamside hornet's nest, and even a few bears.

But only one elephant.

It happened in Zambia in eastern Africa on the Zambezi River. I was fishing for tigerfish—fearsome fanged gamefish with a salmon's adipose fin and a barracuda's toothy, baleful gaze.

Initially, I had hoped to wade and fish the river's inviting shoal waters but was warned that either a croc or hippo would get me. So on a hot, African afternoon I was relegated to casting from a small dock. Just a few minutes earlier, I had hooked what would be my one and only tigerfish for the trip. The fish slammed a spinner in heavy current then leaped throwing the hook almost immediately. Getting a hook-set past those bony, formidable teeth was not easy, it turned out.

The camp where I stayed stood at the foot of an escarpment covered in dry scrub forest filled with game. At night hyenas called from the hills while hippos grunted and wallowed in the shallows like ogres having a day at the beach. Even the deep chuffing of lions could be heard somewhere across the river in famed Mana Pools National Park in neighboring Zimbabwe.

So it shouldn't have been a surprise when three elephants wandered into camp maybe fifty feet from where I stood on the dock. The lead bull turned briefly toward me and threw his trunk up as a warning that he could swat me like a gnat if he felt like it.

I froze, wondering if diving into the river and being eaten by a croc might be better than winding up between an elephant's toenails. But the elephants were not interested in me. Instead, they wandered down the riverbank fifty yards to a tree full of ripe berries.

The big bull, clearly the leader of the group, took charge. It wrapped its trunk around the tree then shook it violently. Thousands of berries rained down around them. Then all three gorged by the trunk-full like kids scooping up candy from a busted piñata.

I watched transfixed by this display, but apparently not out of harm's way. The camp's owner, a transplanted Kenyan colonial, saw my predicament and called to me from the safety of the main lodge up a steep hill about one hundred yards away.

She said in her calm British accent that I really should come toward her as quickly and as quietly as possible. Her tone belied either fear for my well-being or not wanting to deal with the unpleasantness of a squashed guest. Maybe both.

So I stepped off the dock and began walking briskly. I dared not take my eyes off the three elephants, all of which seemed too engrossed in eating berries to bother with something as insignificant as a human carrying a fishing rod.

That is, except for the big bull.

Five tons of elephant suddenly stopped eating and glared at me. Ears bent outward—radar dishes locking onto a target. Warhead armed.

Then it charged.

In a blur of arms, legs, and tackle, I ran harder and faster than I ever had.

African scrub forest whipped past in fast motion, as everything I was told about the running abilities of elephants flashed through my brain. I knew they can run just about as fast as a sprinting human, but can do so for great distances, meaning that catching you is not a matter of if, but when. They also have the nasty habit of picking up pieces of brush and logs with their trunk as they sprint along, throwing them at you, trying to trip you in a final humiliation before you become elephant road-kill. And when they do catch you, the preferred coup-de-grace is to kneel on you, collapsing your rib cage and popping you like a giant thrashing pimple.

But as it turned out the bull stopped after a single step. It was a bluff charge. Not even a charge; more like an elephant's version of a raised eyebrow—almost as if it had turned to its buddies and said: "Hey, wanna see this guy haul ass?"

Of course I was told all of this by the camp owner after I set the quarter mile land-speed record for a man carrying fishing tackle while trying not to soil himself.

And as for the tigerfish . . . wait, what's a tigerfish?

All Hail the Falklands Mullet

There is a coldwater gamefish that prowls along shallow seacoasts and will readily take a well-cast fly. When hooked, it pulls like a bulldog, runs, and jumps. And when you cast for them, you will have wide expanses of shoreline all to yourself. But here's the catch—the best place to hook one is on the Falkland Islands—that relic of British colonialism in the Sub-Antarctic.

Taxonomically, the Falklands mullet swims alone—the only member of its family. And it's not a mullet. Some call it a species of rock cod; others say a type of blenny. It is neither. It superficially resembles a redfish with an underslung mouth but is leaner with long pectoral fins like a jack. It is as isolated as its habitat.

I was trying for sea trout when I happened into them. The remote Falklands are best known for the short war with Argentina in 1982. As a byproduct of the ten-week conflict, hundreds of acres of minefields left behind have become a refuge for penguins, and five species nest there. Courtesy of long-ago stockings from the British, sea-run brown trout are found in most of its rivers. I had heard vaguely that mullet could also be caught but mostly by bottom fishing—pieces of mutton were frequently mentioned as the bait of choice in this sheep-rich country. Compared to the more glamorous sea trout, Falklands mullet sounded like a trash fish.

Armed with a fly rod, I traveled from the quaint capital of Stanley—the only town of any consequence in the Falklands—to Moody Brook a few miles away and the site of the first sea trout stocking in the 1940s.

But when I arrived, Moody Brook was apparently in a bad mood that day, running pitifully low and unimpressive through a treeless landscape. It drained through a small culvert under a road. The brook seemed an unlikely place to catch a sea trout—or anything for that matter. So instead, I looked to where it trickled over some rocks before emptying into Stanley Harbor.

Then I thought I saw something pushing water in the shallows—a sea trout? So I strung up my fly rod and clambered my way to the shoreline over slick cobbles. The water was clear and cold with a craggy bottom of rocks and rubble.

I cast a Deceiver—a standard searching fly—but caught nothing even though I could still see the subtle wakes of several fish moving around. So I switched to a dark, weighted Clouser Minnow and let it sink before twitching in among the rocks like a crab or shrimp.

Something grabbed the fly with a sharp tug. I set the hook by pulling down hard on the line. The fish boiled in less than a foot of water then powered deeper into the harbor taking line in surges.

Eventually, it turned and I could see it was no sea trout. It had drab, dark brown sides and a flat, wide head—a Falklands mullet. Then it tore away one last time and remarkably ended its run with a vertical jump that would make any sea-run brown proud.

I beached it, about a three-pounder, and finally got a good look at it. It was plain, but not ugly. Its coloring was on the somber side, which matched its rugged surroundings. It was neither streamlined nor blockish. The word pragmatic came to mind—a fish that needed no fancy spots, graceful lines, or extra bobbles to get its job done. I released it into the shallows, where it pushed a wake before heading into deeper water.

I wound up catching two more over the next hour. By the time I landed the last fish, a young couple had stopped along the road and began to talk to me. The man turned out to be a soldier on a weekend leave from the military base in Stanley. I offered the mullet to him, which he happily took and put in the trunk of his car. He assured me they were excellent eating.

Back in Stanley that night, I stopped at a pub and was delighted to see mullet on the menu. I ordered it broiled and it came out perfect: delicate yet hearty, kind of like the Falklands themselves. A pint of Guinness was the perfect accompaniment.

This fish had it all, which got me thinking:

Just the same way brown trout were imported to the Falklands and throughout South America by the trout-obsessed British, perhaps mullet-lovers could bring this shallow water gamefish to the north. I could think of a few coldwater coastlines that could use a shallow water fish that takes flies selectively, runs and jumps, and is flat-out delicious.

But then would come mullet tournaments, mullet guides, mullet snobs, mullet specialty fly lines, not to mention mullet seasons, quotas, and regulations. Commercial and recreational fishermen would argue over their management. "Mullet wars" would break out.

Sipping my pint and enjoying another forkful of this toothsome, mysterious gamefish, I thought better of it. Maybe the Falklands mullet is best where it belongs—on its namesake remote outpost in the lonely South Atlantic.

Finding Oscar

All anglers dream of casting somewhere that has never, ever been fished before: a virgin river, lake, or coastline where the fish are large and naïve and unsoiled by a single bait, lure, or fly. Few of these places exist anymore—and with more than seven billion of us wandering the planet, their numbers, sadly, will only diminish.

Once—just once—I got to fish such a spot—a forty-acre lake just a short hike through a small woodland better known as the Amazon rainforest. The lake had no name and was one of thousands left behind when the mighty Amazon River ebbs some forty feet in the dry season and every pothole and low spot suddenly becomes a real-life fish bowl.

This particular lake also contained a population of sideneck turtles—a protected species—and I was able to tag along with a team of Brazilian researchers studying them one morning. When I got there, some of the team had just pulled in a net they had set out earlier. In it were several turtles, which they tagged and released.

A large fish wound up in the net, too—a piracucu that had become entangled and died. It weighed about twenty-five pounds—to me a giant, but in reality a mere baby after I learned they can weigh four hundred pounds and reach fifteen feet in length. One of the researchers took a machete and hacked off long fillets, while his friend made a campfire. When it burned down to coals, they tossed in the fish skin side down where it sizzled and smoked. Two armored catfish that also wound up in the net were thrown in the fire whole.

When the piracucu was done cooking, it was laid out on a log and we helped ourselves. It flaked into meaty chunks and was very good. The catfish were cracked open and the meat scooped out like cooked crab.

It was over this unusual shore lunch I learned that except for the few times something accidentally wound up in the turtle nets, the lake had never been fished before. It was too small for commercial fishermen to bother with, and there was virtually no sport fishing done in this remote corner of the Amazon. When I mentioned I had a pack rod with me, one of the researchers agreed to stay behind after recording his data for the day and take me out in a canoe to fish.

An hour later, we shoved off in a classic jungle craft—a six-foot dugout with no more than an inch of freeboard. I sat precariously on the fantail while my newfound fishing guide paddled from the bow.

As we headed farther from the shoreline, the lake's entirety came into view. It was ringed in thick, dark jungle. Baitfish occasionally scattered here and there as unseen predators chased them. The raspy calls of black howler monkeys could be heard off in the distance. Overhead, a toucan flew by, its Technicolor bill glowing in the Amazonian sun.

I made a long cast with a large inline spinner, which seemed like a good searching lure for the lake's dark waters. I let it sink then jerked the rod to get the blades to turn. Now I could feel the low thrum of the spinner working—the first lure ever—*ever*—retrieved through these mysterious waters.

With every turn of the reel handle, I tensed knowing it would only be a few moments before . . . WHAM! Something grabbed the lure violently and was immediately hooked. The fish grinded out a few feet of line from the reel then jabbed repeatedly for the bottom.

What could it be, this first fish ever caught by a rod and reel in this virgin lake? My mind raced. There are thousands of freshwater fish species found in the Amazon Basin. Which one was it? Could it be a fearsome piranha? There are as many as thirty varieties alone. How about a catfish? An astonishing twelve hundred species live in the Amazon ranging from the massive nine-foot-long piriaba, to the diminutive—but horrifying—candiru, a parasitic species that purportedly swims up human urinary tracts mistaking them for the gills of their usual host fish. Oh God, please don't make it a candiru.

What about an electric eel, all 650 volts of one? How about a freshwater stingray? Maybe it was the pirarucu's brother?

I pumped the fish closer and now saw flashes of color—deep oranges and reds. It wasn't big, but it didn't matter. Whatever it was, it clearly would be an exotic, exciting new species for me, something worthy of living in a lake that had never been fished before.

Not exactly, it turned out. For when the fish finally yielded and came to the surface, not only had I already seen it before, I actually had one at home living in a fish tank. It was an Oscar about twelve inches long and weighing maybe a pound, the exact same size as my pet, named unimaginatively "Oscar."

If you've never heard of an Oscar, go to your local pet store and stroll down the tropical fish aisle. You will find them next to the guppies and neon tetras. Cute little things and you can get a small one for about twelve bucks.

I grabbed the fish behind its head, now admittedly feeling strangely guilty that I had hooked my pet's relative. I removed the spinner and showed the fish to the researcher who considered it for a few moments then said I should let it go. I placed it in the warm, dark water and watched it swim away—the first fish ever caught and released in this nameless lake.

Pet or no pet, lucky for the Oscar we had already eaten.

The Cuban Eight-Weight Crisis

Breaking a fly rod while surrounded by tailing bonefish is bad, but having it happen in Cuba, where the nearest tackle shop is not only in another country—but across a political divide far wider than the mere Straits of Florida—is much worse.

There are no Orvis shops in Cuba. You can't send an email to Winston or Sage and have them whisk you a new rod overnight via FedEx. In fact, there isn't much in the way of any material goods or selection the way we might expect it in the capitalist United States. For example, Cuba has one brand of beer—Bucanero. Comes in either light or dark. I recommend the dark. Another example was my guide, a very knowledgeable angler named Raidel Almeida, who didn't own his own fishing rod.

And so on this February day in a tiny skiff at Las Salinas—a massive network of tidal flats along the westernmost shores of the infamous Bay of Pigs—my rod broke, and I had no backup.

Up to that point, it had been a delightful morning. I had just driven down a thirty-mile-long dirt road with a group of American and Cuban conservationists on a scientific exchange. We marveled at the bird life in this surreal landscape. A riot of waterbirds—fuchsia Caribbean flamingos, roseate spoonbills, massive white pelicans—crowded on salt flats along with innumerable lesser shorebirds. Beyond them in the mangroves, warblers flitted about readying for their northern migration. A month later, they would fly into the United States oblivious of embargoes and blockades.

At the road's end stood a small research station—a cinderblock compound that to my delight also housed a single tourist amenity: a few skiffs and guides who could take you bonefishing. I had heard about this angling oasis beforehand and had brought along a four-piece fly rod and box of flies in my backpack just in case. After a quick negotiation with the group, I bid goodbye to my ornithologist friends for a few hours. Raidel and I climbed into the motorless skiff and began poling into Cuba's tidal wilderness.

I had dreamed of fishing in Cuba ever since I had read *Salt Water Fishing* by Van Campen Heilner—a classic, time-capsule of a book that captures a brief period in the early twentieth century of near-virgin angling from the tropics to British Columbia. His stories about Cuba are particularly fascinating, with tales of expeditions to jungle rivers where you could walk on the backs of hundred-pound tarpon—or so he writes.

Then the revolution came and much of Cuba's coastline was never developed for tourism—unlike the rest of the resort-choked Caribbean. Earlier that morning, bouncing down the dirt road as we drove deeper into Las Salinas, I kept thinking this is what the Florida Keys must have been like sixty years ago or more. This felt especially true when the lone car we saw during the entire drive rumbled past. It was a pre-revolution relic—a faded royal blue 1955 Oldsmobile.

About a mile onto the flats, Raidel stopped poling the skiff and told me there were bonefish ahead. When I excitedly asked where, he cautioned me to speak quietly in interesting broken English by whispering "shut up" over and over. And I did.

Now I could spot the bonefish—a pod of maybe half a dozen shadows on a sand-colored bottom making their way toward us from seventy feet away. I stood at the bow laboring to get the fly out into a stiff breeze quartering over my right shoulder. But before I could lay out a decent cast, they spooked. So I reeled up my line, and we headed farther onto the flats.

A few minutes later, Raidel spotted another small school tailing just off an islet of mangroves. And that's when it happened. I'm not sure if during my earlier casting attempt the weighted fly nicked the rod blank, but as I flexed the fly rod to get the leader through the guides, it made a sickening snap two feet from the tip.

I cried out, having suddenly been transported from Las Salinas to Dante's Inferno. Raidel asked what happened and I pointed to the rod showing the tip of my eight-weight now dangling by a few splinters of graphite. I slumped down in the little skiff, crushed. My fly rod was broken. My day was over.

Or was it?

As I sat and pouted over my bad luck, the man who owned no fly rod simply twisted off the broken tip, restrung the rod, and handed it back to me.

"Cast," he said, pointing to just off the front of the boat. "Bonefish."

During the past few days in Cuba, I had come to learn that Cubans were tough, resilient people. You can see it most obviously in their American cars that somehow still run despite having no replacement parts since 1960. So they manufacture their own or improvise. In a taxi in Havana, the cabdriver kept the broken passenger door handle of his '53 Chevy in his glove box, which he handed to me when we arrived at our destination. I fit it into the hole in the door, opened it, and then gave the handle back to him along with my fare.

This attitude is known among Cubans as "*A lo Cubano*," which translates to "the Cuban way"—an unwritten rule that Cubans will persevere no matter what the circumstances. You often hear Cubans say this to each other with a certain knowing sense of pride. Regardless of one's politics, there is a lesson to be learned here.

I stood at the bow of the boat once more. The eight-weight now felt more like a stubby ten-weight. Grunting into the wind, I managed possibly the ugliest forty-foot fly cast ever made. Yet somehow the line unrolled and the fly dropped four feet in front of a bonefish, which immediately swam over and inhaled it.

I set the hook, and the fish streaked across the flat sending my reel into a blur. It swam around the boat a few times surging here and there before tiring. It was small by bonefish standards, maybe two pounds. But given the last few minutes, I couldn't have been happier if it weighed twenty.

I leaned over the boat's gunnels, cradled the bonefish, then turned to show it to Raidel who was grinning proudly. He knew what he had done for me.

Then he said in his broken English: "You catch that fish *A lo Cubano*."

Now it was my turn to beam, having just received the greatest compliment on this perfect day of bonefishing on the island of Cuba.

Forbidden Love in Guatemala

The fish streaked through blue water, crashed the trolled bait so hard I thought the rod would snap in half, and then leaped impossibly high and far—all in two seconds.

I had just witnessed my first hooked mahi mahi.

My friend Richard held on to the rod while the fish continued running and jumping and bounding over waves. Line poured off the light bait-casting rig he brought along for extra sport.

He got his money's worth as the fish now plunged deeply under the boat taking another fifty yards of line, then changed its mind and cleared the water twice more.

Eventually, the fifteen-pound mahi was gaffed and hauled over the gunwale with its electric colors on full display—a Jackson Pollock painting of splattered blues, greens, and yellows.

I stood starstruck. It's not every day you meet the aquatic version of Superman in a Technicolor dreamcoat.

Too bad it was just bycatch unceremoniously dropped into the fish box where it would later become our lunch. You see, this was a sailfish trip, out of a sailfish lodge in Guatemala that prided itself on the number of sailfish it raised, hooked, and released. Period.

Less than an hour earlier, I had technically just caught the largest fish of my life—an eight-foot-long Pacific sailfish that weighed nearly one hundred pounds. I say technically because I had little to do with catching it. A trolling rod went off, which was then thrust into my hands by one of the mates. I grabbed it while a purple javelin of a fish tail-walked off in the distance. Then the captain started screaming at me: "REEL! REEL! REEL! REEL!" as he backed the boat down while the sailfish ran.

I cranked like a crazy person gobbling up slack line as quickly as I could and wondering if the fish was still there. It was. No more than five

Guatemala Mahi Mahi

minutes later the sailfish was at the transom. One of the mates grabbed the leader—making it a fair catch in the big-game rulebook—then broke the line on purpose. This was followed with lots of pats on my back and an overly enthusiastic "NICE CATCH!" by the captain.

What the hell did I do?

Meanwhile, no high fives for the mahi that was dying in the cooler.

Even before the sailfish and the mahi, I could sense this might not be the trip for me. I was invited by a group of friends who needed a fourth to round out their party so I signed on. I had never caught a billfish—let alone stayed at a fishing lodge that specializes in catching them. Little did I know how seriously this last part was taken.

The first tip-off came just a few miles out of the inlet when we came across acres of five-to-six-pound Pacific bonito tearing into baitfish on the surface. I excitedly readied the spinning rod I had brought along and waited for the boat to slow so we could start casting. If the bonito were anything like their Atlantic counterparts, I was in for a great light tackle fight. But we never even slowed down. You see, they weren't sailfish.

I watched as we passed the massive school at thirty-five knots, feeling like the proverbial kid pressed up against the candy store window with the door locked.

Don't get me wrong—I understand that billfish anglers are a focused group, the same way striped bass, salmon, and trout anglers are. When we finally slowed to start fishing, I didn't mind monotonously trolling six rigged baits and lures behind the boat either—the usual complaint for this type of fishing. But whenever we hooked a sailfish, it was more of the same: chasing it down at high speeds, then a quick release. It got a little boring.

Eventually we came to a large dead tree bobbing in the gentle Pacific swell. As we trolled past, I looked in the shadows below it and could see grayish shapes suspended in shafts of sunlight. One of them sprinted past the boat, slammed a trolling lure, and began tail-walking. Another mahi. One of the guys grabbed the rod and fought the fish as it continued going through its ferocious display.

I heard that a hooked mahi is sometimes followed by other members of its school, so I grabbed my spinning rod again and tied on a bucktail. I cast near the hooked fish and began twitching the lure back to the boat. Something ghosted behind it, so I sped up my retrieve. The fish accelerated and hit hard. I had hooked my first mahi. Both fish turned out to be less than ten pounds—but they fought incredibly hard. And clearly there were more under the floating tree.

After the mahi were landed, the captain said something about how there were no sailfish here and started motoring away.

I couldn't take it anymore.

Feeling like Oliver Twist asking for more gruel, I said: "Any chance we could stop and cast a few times?"

The captain gave me a look I had seen once before. It was while fishing for Atlantic salmon in Nova Scotia. After a fishless morning, I stood on a high bank and spotted a few striped bass in the ten-pound class holding in a pool along with some indifferent Atlantics. Two salmon fishermen methodically worked the pool with dry flies. I called across the river to them, pointing to the stripers and suggesting they might want to switch to a streamer and a sinking line since the salmon were not hitting.

They gave me the same look the captain was now giving me. It's the look that says: "It's simply not done here."

And it's not, at least not in Guatemala in a lodge that boasts of the numbers of sailfish raised, hooked, and released. So the big sportfisherman kept trolling its pattern of six lures and baits farther and farther away from a school of mahi and an afternoon of fun. All I could do was look at the floating tree, its salt-bleached limbs bobbing back and forth in the boat's wake as if waving a long, sad goodbye.

Ten Casts in the Congo

The goliath tigerfish weighs up to 120 pounds and is known to occasionally attack people. Purportedly, it mistakes flashing jewelry for baitfish and in the process maims arms and legs. Its exaggerated fangs are almost cartoonish, drawn by a mischievous eight-year-old with a penchant for monsters or velociraptors. Its deeply forked tail screams speed, and the few anglers who have hooked into one report that it jumps like a crazed tarpon.

I wanted to catch a goliath tigerfish badly.

Just once, I had a brief chance when I wound up in the Republic of Congo in Central Africa on a work assignment. I was staying for a few days at a research camp on the Sangha River, a large tributary of the massive Congo River—second only to the Amazon in sheer watery volume. To get there, you had to fly on an ancient Russian commercial jet from Brazzaville to the river town of Ouesso, take a boat twenty miles upriver, then lumber over miles of pitted logging roads via four-wheel drive.

The camp itself was surprisingly comfortable with a diesel generator, air conditioning, and hot showers. Next to it was a small village, and beyond that loomed the heart of the Congo Basin Rainforest—an Alaska-sized domain of lowland gorillas, forest elephants, leopards, and other lesser-known creatures like sitatunga and bongo.

The village was a true frontier town. A generation ago, it was a remote settlement of subsistence hunters and fishing families. Now, an overseas logging company was the main employer, and a massive sawmill rumbled off in the distance. Many of the local people walked around with a bewildered look as they witnessed their traditional way of life vanishing before their eyes.

One morning I got up early, unlocked the back door, and walked past the armed guard who patrolled the research camp. Leaving the air conditioning and entering the dense African air felt like being draped in a hot, wet blanket.

A trail led down a grassy hill to the river. Wooden boats were hauled up on the muddy bank, and a few fishermen mended their nets. The river made a sharp bend here, and a wide pool of confused currents and backwaters swirled and churned before gathering for their eventual confluence with the mighty Congo. An earthy, fertile smell wafted from the dark waters, while tree branches, palm fronds, and other flotsam drifted with the current.

I took my three-piece pack rod out of its tube and assembled it, clamping the reel to the seat then stringing line through the guides. Even though the sun was just peaking over the trees on the far side of the river, I was already sweating. I opened my tackle bag and chose a large bright plastic popper. It seemed like a good lure to entice a goliath tigerfish into striking.

Two fishermen stopped what they were doing and walked behind me to watch as I made my first cast. I turned to them, smiled, and bid them a hearty *bonjour*—one of just a handful of French words I knew. They nodded but did not smile back. A few fishless casts later, I decided to switch to a large metal spoon. By the time I retied and began casting again, a few more people had gathered to watch—or was it to stare? Perhaps they had never seen anyone fish with a rod and reel before—a definite possibility in this remote jungle outpost.

Quickly growing more impatient with the fishing, and maybe feeling a little uneasy, I switched lures again, this time to a diving plug. By now the group of watchers had swelled to a dozen—a mix of men, women, and kids.

And still, no one smiled—they stood there stoic and silent while I continued to cast. And then it occurred to me: Why should they smile? A few years earlier, the country had gone through a civil war. Before that, there was nearly a century of brutal colonial rule followed by two decades of failed Marxism and associated coups and assassinations. Now a timber company was swallowing up, log by log, their centuries-old way of life.

And now here I was with my high-tech three-piece graphite rod and expensive lures that probably cost more than entire families earn in a year. And I was casting for a fish that if I was lucky enough to catch, I planned on releasing. I might as well have been the proverbial tycoon lighting cigars with hundred-dollar bills.

And so, with a growing crowd of unsmiling Congolese collecting behind me, I reeled up my line and hung my lure in its keeper ring. Then I walked past them nodding respectfully before heading up the trail and leaving the goliath tigerfish to those who needed it more.

Part VII

LATER, GRACELESS CASTS

Hickory Shad

The 20/40 Club

One September afternoon on Block Island, my friend Jim Turek squinted and strained as he tried to thread a leader through the guides of his eight-weight. Eventually, he gave in and fumbled for reading glasses hanging from around his neck. Now he took on the demeanor of a librarian looking up a Dewey Decimal number rather than a rugged saltwater fly fisherman readying his gear.

I guess he could see the amused expression on my face when he said: "Yeah, my eyes just went. Happens to everyone when they're in their mid-forties. It'll happen to you before you know it."

Not me, I thought. I'll never get that old. I had seen those pages in fishing catalogs featuring strange fly boxes that help you pre-thread your flies. You'll find them along with wading staffs and those waders with waterproof zipper flies. I always breezed past these geriatric gadgets with an air of smugness while looking for, you know, serious fishing gear.

A few years later on the annual Delaware River camping trip, I bade farewell to my non-angling friends lounging around an early evening campfire. With a belly full of hearty camp food, I strode down a forested trail in my waders. The sun was getting lower and the calls of woodland birds—metallic rattles of veeries and sad, fluty notes of hermit thrushes—echoed from the forest around me.

The trail ended at a favorite pool. Here sulfur-colored mayflies would hatch just before dark, and trout would rise to them. It was the sweet of the evening and I relished it.

I waded into the river just as the sun dipped below the woods behind me. On cue, the first trout rose just twenty-five feet from me. I waited and let it come up a second time, then a third before I stripped off line to make a cast. Then downstream, another trout started working. Though the first rise was closer, the second fish might be easier to hook with a simple downstream drift. I decided to fish for it first. After I landed it, I told myself, I would go back and catch the closer one. By then, there should be other targets within range. This could be a night of a half dozen trout landed or more.

The fly line shot out as I checked the rod upstream. The leader fell in S-curves a few feet above the fish. The fly touched down directly in its feeding lane. Now it was just a matter of feeding slack into the cast and letting the fly drift into a waiting maw.

The trout promptly came up—and took a natural right next to my fly. Hmmm. I picked up the line and cast again with the same result. No matter. I reeled up the fly line, matter-of-factly snipped off the sulfur dry, and grabbed a snowshoe emerger from the dry patch of my vest. This is the go-to sulfur pattern when trout are even the slightest bit selective. Just tie one on, rub in some floatant, and hang on.

More trout—some clearly very nice fish—were now casually sucking down sulfurs all around me. I held the end of the tippet in one hand and the emerger in the other, then went to thread it through the eye of the hook. Wait a second. Where is the eye of the hook? In the fading light, all I could see was some blurry yellow thing. I blinked and shook my head a few times.

Now a trout rose fifteen feet from me just begging for a snowshoe emerger. I tried to thread the fly again with the same result. Better click on my headlamp; must be darker than I thought. Now I stared at a thoroughly illuminated—but just as blurry—yellow blob.

Trout continued to rise all around me. I alternately held the fly as far away as I could with outstretched arms, then just inches from my straining eyes. Same result. By now I might have let out an anguished whimper.

With the river now virtually boiling with trout, I stood there for what seemed like hours desperately trying to accomplish this most basic and fundamental fishing task. I failed.

It was no use. I cranked up my line and leader all the way through the guides and onto the reel. I was done. I stepped out of the river to the sounds of unmolested trout gorging behind me.

The forested trail seemed darker than normal. I trudged along periodically shaking my head in bewilderment at what just happened. Back at camp, I slumped in my chair next to Paul who was sipping a beer and enjoying the fire. He had just returned from upstream and reported that he had a few good shots at rising fish but couldn't get any. At least he had made more than two casts.

Then he asked me how I did, so I told him what happened.

Paul, who is a few years older than me, nodded knowingly and reached into his fishing vest. He handed me a zip-locked bag with a pair of magnifiers that attach to the brim of your cap.

"They're called Flip-Focals," he told me. "Keep 'em. You're gonna need them—forever."

I slipped them in my vest pocket next to my dry fly box and floatant—two things I would never dare be without while trout fishing a sulfur hatch in late May. Now there were three things.

If you don't have your own pair yet, they're under Tools and Accessories on Orvis's website. Better get them soon.

Because Jim Turek is right, the bastard.

Respect the Hex

There is a common angling superstition against bananas, which some say bring bad luck. Some charter captains go so far as to post signs on their boats stating that no bananas are allowed on board. Apparently this hex has its origins in Hawaii. There, bananas were banned on long-range fishing trips because they spoil easily and cause other fruits and vegetables to rot prematurely. How this mutated into a fishing jinx, no one knows.

Frankly, the banana jinx is stupid. I have eaten bananas many times on fishing trips with no ill effect on the day's catch. File this silly superstition with black cats and stepping on cracks on the sidewalk.

But there is a real angling hex and it has laid waste to otherwise fine fishing days for me and other True Believers. Here's how it goes: There are Fishing Gods and they can be mean and spiteful. They owe you nothing—no matter how much time may have "paid your dues" or how great the day's conditions might be. Nothing makes them angrier than hubris, such as guaranteeing you will bring home dinner, or stating with absolute certainty that you are going to fill the boat with fish.

Things like buying a lemon before a fishing trip or filling your cooler with ice are particularly bad.

So whatever you do, don't piss them off.

I have learned this from painful firsthand experience. I began noticing the existence of a hex during my party boat bluefishing days. It would happen any time I took advance orders for fillets from neighbors and friends. That's when I would inevitably get skunked, or get seasick, or worst of all, come home with a single two-pounder that faced the distinct possibility of being divvied up eight ways. Yet, if I boarded the boat quietly and fished with humility, I would often bring home a burlap bag overflowing with fish.

Once I realized the mighty power of the hex, I changed my ways. If someone asked me for fish before I actually went fishing, I'd mutter something unintelligible then wander away as quickly as possible. I'd commit to nothing.

Sometimes these same people would say cheerfully: "Catch a lot of fish!"

I'd reply in a dour: "I hope so, but you never know."

But others are more ignorant than me. Pity them.

In Alaska, a fishing guide gave a pep talk the first night of a seven-day float trip about how to release all of the huge rainbows he assured us we would start catching the next day. I held my head and shuddered over what he had unknowingly unleashed. That night and into the next day three inches of rain fell and blew the river out for the rest of the trip as I knew it would.

In Rhode Island, my friend Bob insists on jinxing nearly every fishing trip I've ever been on with him. He starts out when the boat is pulling away from the dock. "I just feel it," he says, "Today we're going to kill 'em."

That's when I attempt to appease the Fishing Gods by saying my usual "I hope so," extra loud. Then I go on about how conditions might not be so great and it will be good just to get out on the water and have a good time. Sort of a reverse psychology counter-hex.

But Bob always shrugs me off. "I'm an optimist," he likes to say as he continues with his glowing prediction of the day's fishing. "Better go sharpen my fillet knife!" Yes, and he's often a fishless optimist—at least when I'm there.

Occasionally circumstances have caused me to let the hex slip my mind, as it did one day on the Delaware River when I had already caught more than thirty stripers on my fly rod. How, you may ask, could you have possibly been hexed on a thirty-fish day?

Here's how: First, I need to point out that the bass were yearlings—eight-to-ten-inchers. I found them while prospecting around some riffled water looking for smallmouth with my six-weight. Still, they were fun to catch, hitting hard then making mini surges for the bottom.

I hooked the first few drifting a weighted streamer through some broken water below a set of rapids. Then I came to a large submerged boulder and found the motherlode. Now every time the streamer swung through the eddy below the boulder a bass would grab the fly. I landed a dozen bass in a dozen casts, then two dozen. I couldn't miss; I was Ted Williams taking batting practice, Michael Jordan at the foul line.

Just as I was releasing yet another striper, my friend who was fishing upriver came up behind me on a gravel bar. Then he called out and asked if I was having any luck.

Clearly the bliss of catching so many consecutive stripers had clouded my judgment because I found myself announcing: "Watch this!" and made another cast. Blasphemy. See above note about hubris.

The fly drifted through the eddy just like it had just done thirty times. Not a touch. I cast again with the same result.

I turned to my friend and explained how I had just been catching stripers on every cast. There was desperation in my voice. He looked at me sharply—he's another True Believer—and we both nodded knowingly at my sacrilege. A few frantic casts later, I did land another striper and turned to show him. But he was already heading upriver, probably trying to stay as far away from me as possible. After that, the fishing quickly slowed before shutting down completely.

A small slap, but point well taken. My apologies; it won't happen again.

John's Haunted Crappie

Ice fishing suffers from an image problem. Among non-fishermen and even a surprising number of otherwise knowledgeable anglers, the main complaint is that it looks boring. They imagine someone hunched over for untold hours in the middle of a snowy, windswept lake. Upon further inspection, the angler turns out to be frozen Jack Nicholson in *The Shining* complete with a long icicle of mucous hanging from his nose.

But ice fishing—at least the way a group of my friends practice it—can be every bit as exciting as casting a dry fly to a rising trout or throwing a popper into a creamy surf for stripers.

This is because we are proud practitioners of the high art of jigging for panfish—perch, bluegills, and crappie. We use rods that look like they came from GI Joe's footlocker—a two-footer is a long one. Scale-model reels hold less than a hundred yards of line. Four-pound test is as heavy as you go; two-pound is better. For lures we use ice jigs—chrome-plated or painted lead-heads. Half a dozen can fit on a nickel. We sweeten these with insect larvae; waxworms or mousies—a type of maggot—work best.

And then, using hand augers that resemble giant corkscrews, we drill into the ice and scoop out the shavings to reveal a mysterious black hole that is nothing less than a portal into another world. What lurks down there? To find out, you lower the jig and watch coils of line vanish into the darkness. When they finally stop, you reel up a turn or two and then ever-so-gently jiggle the rod. You stop and study your line or rod tip. Nothing, so you jiggle again. Stop. Then you jiggle again but notice an ever-so-slight change in the way the jig feels. It's not a hit; it's a hunch. You lift and the tiny rod lurches into a graphite divining rod pointing downward, connected to an unseen fish. Maybe a few feet of line spills off the miniature reel, the drag ratcheting like an unwinding pocket watch.

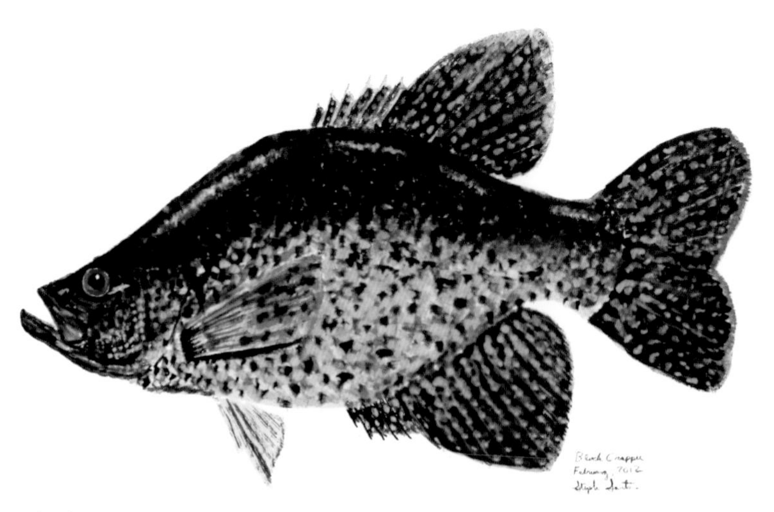

Black Crappie

Eventually the fish begins to yield to the rod's pressure, and you gain a few turns of line. Then you see your first flash of color. Was that a green back and a flash of orange? Then a head appears followed by the humped back, vertical bars, and the brilliant orange pectoral fins of a nice yellow perch. With gloved hands, you swat it out of the hole grizzly bear–style. It lays on the ice with its impressive dorsal spines erect and sharp gill plates flared. If you were a raccoon or heron this might give pause. But since you are neither, you slide the perch a few feet away where it will flop for a minute or two before settling down. Later, with several of its schoolmates filleted and rolled in Japanese bread crumbs then deep-fried and washed down with good beer, it will make a cold winter day seem downright luxurious.

But it's not that easy. You see, panfish schools move around, so this means you have to keep drilling holes to find them. You might get two fish from hole number one, three from hole number two next, and zero from the next half dozen. So you drill some more until you hit the motherlode—half a dozen foot-long perch and a few big bluegills come out of the hole as fast as you can lower your jig. Then you drill some more. Over the course of a few hours, you might drill thirty holes. The next day, your shoulder and arms will ache.

Catching all these small fish sounds boring, one might say. Perch, sunnies, crappie . . . who cares?

Right.

One day I was ice fishing with my friend John Waldman, a writer and fisheries biologist who first introduced me to panfish jigging. He fishes a series of reservoirs that provide water to New York City. Each one is different: Some are shallow and known for great numbers of sunfish; others are deep and contain large brown trout. The one we fished that day was unusual because it supports schools of white perch, a smaller cousin of striped bass that can live in both fresh- and saltwater. When this particular reservoir was created in the late 1800s, some perch, which had migrated from the tidal section of the Hudson River, became trapped and have since flourished as a landlocked population.

White perch are the most mercurial of the panfish species; they roam the reservoirs like pelagic tuna endlessly searching for schools of baitfish. One day you can catch them in a weedy cove in two feet of water, the next day find them suspended forty feet off the bottom of a one-hundred-foot deep hole. Most times, you don't find any. But when you do, you can catch them by the dozen. And they are particularly good to eat.

We began drilling holes in the deepest part of the reservoir gradually working our way into shallower water searching for schools. I had brought along a brand new Finnish auger with surgically sharpened blades that could slice a four-and-a-half-inch-wide hole through a foot of ice in mere seconds. John on the other hand had dragged out a beastly old auger with eight-inch blades that needed sharpening. For every hole he drilled, I drilled three, which allowed me to cover far more water.

I was one hundred yards from John when I hooked and landed my first white perch. Then I briefly hooked a second fish that I lost. Perch schools can move quickly, so I called to John that I found some fish and he should come over before they leave. He walked over and was about to drill a hole with his massive auger, but I told him he should use one of the extra holes I had just drilled. So he did and lowered his jig to the bottom.

Almost immediately, he had a subtle strike that he missed, followed by another that he hooked. The rod bent deeply but did little else. At first, we thought he had snagged into one of the dense weedbeds found in the reservoir's shallower water. But then the weedbed began to move, sluggishly pulling the rod downward in slow surges. This was clearly a nicer fish, so I reeled up my line to watch. John's rod never really stopped surging, but slowly he craned the fish closer to the hole. By now I was standing next to him, transfixed on the black portal.

Then we both saw an enormous open mouth with the ice jig caught in it. The mouth itself could fit in the hole, but the rest of the fish couldn't. This eliminated a big bass, whose gaping maw is often the widest part of the fish. Indeed, I had pulled bass through holes cut with my Finnish auger that weighed nearly four pounds.

We then realized we were looking down the mouth of the largest crappie either of us had ever seen. You sometimes come across taxidermy mounts of these beasts hanging in old tackle shops. The fish look impossibly huge—big around as a pizza pan and weighing three or even four pounds. Their wide speckled fins and large dark eyes make them look like some wondrous deep-sea creature rather than simply just another panfish.

But try as he could, John could not get this crappie-of-a-lifetime to fit through the hole.

Desperate, he reached into the icy water and grabbed the fish by the lower jaw. It was his now, if only we could figure out how to extract it through the ice. Then I got an idea. Maybe I could drill another hole right next to the first one, making a sort of figure eight that the crappie would fit through. This seemed like a good plan, until we realized John's hand would be maybe an inch away from extremely sharp spinning auger blades. Losing a finger—or worse the giant crappie—seemed too risky.

How about if we could get a rope down the fish's mouth and out its gill? Then we could hold it there as long as we wanted while we leisurely expanded the hole. I rummaged through my tackle and found a bungee cord that could be modified into a rope.

But human physiology was catching up to us. All this time, John had his hand submerged in the 33-degree water up to his wrist. He called out in the voice of a man who was losing his grip—both literally and figuratively: "Can't . . . hold . . . on . . . hand . . . won't . . . work."

Meanwhile, the crappie, which had remained docile during most of the ordeal, suddenly lurched its head, pulling from John's frozen fingers and throwing the jig in the process. We watched—horrified—as the huge mouth, still gaping, sank into the hole and vanished.

Forever.

A great groan echoed across the reservoir. John dropped to his knees. I lay prostrate on the ice. It was a crushing defeat for both of us.

We remained still for a long time, listening to the quiet that is a frozen lake in February. Eventually we gathered ourselves and continued fishing. But our hearts weren't in it. Plus the perch school had long since moved on. Eventually we left.

To this day, if ever I want to make John wince, all I need to say is, "Man, that was a big crappie."

Too bad ice fishing is so boring.

A Wild Dilemma

Thy rod and thy chain stringer—they comfort me. Blame it on my electric smoker, but I am an unabashed fish hog when it comes to hatchery trout—especially in heavily stocked waters. Each spring, I unleash myself upon marginal trout streams with a jangly aluminum chain stringer stuffed in my fishing vest next to boxes of lovingly tied barbless flies normally reserved for Sacred Wild Trout.

For their hatchery brethren, dainty dries are generally skipped; instead I fish no-nonsense stuff—big Woolly Buggers, droppers of gaudy wet flies, anything that I think will fool a fresh-from-the-hatchery stocker into striking.

And if they do strike, I pounce like an osprey, my own talons mercilessly finding their way into gills; or I haul the hooked fish mackerel-style onto the bank as dirt and leaves stick to their slime coat. Then I clasp it to the Jaws of Death, namely the $1.99 stringer made in China.

Though I get no pleasure in taking the life of any fish, there is some inner hunter/gatherer that thoroughly enjoys dragging around a full stringer of dead trout. It's hard poundage—or better yet hard calories. And I enjoy gutting fish at streamside, washing the body cavity with river water, and watching entrails float away where they will be eaten by crayfish, or aquatic insects, or schools of minnows.

And I really like hatchery trout smoked in applewood. So do my friends who seem to universally cherish this delicacy when I dole it out to them. I know few people who will say: "No thanks; I don't want a smoked trout."

Lastly, I love the schedules state fish and wildlife agencies post for stocking trout—often with exact date, location, and number of fish to be stocked. Make no mistake, hatchery trout are put in for you to catch and take home. Don't worry; they'll make more.

But sometimes things don't go as planned even when fishing for normally predictable stockers.

I was fishing the Neversink River in the Catskills: not the hallowed Neversink Gorge Unique Area known for its wild fish, amazing scenery, and rigid catch-and-release policy, and not the utterly posted water above the reservoir where Theodore Gordon, the father of Catskill fly fishing, once cast.

Instead I fished the lower Neversink in a town called Cuddebackville. This is a workingman's stretch of river, heavily stocked and becoming bathtub warm come late June. But this was late April, and the river still had the look and feel of a fine trout stream with a good head of cold, tannin-stained water. Even a few caddis and mayflies were fluttering about when I strung up my rod around 6:30 in the evening.

I came to a nice trouty run where two fishermen were already working the far bank. Both fly fished, though one also had a spinning rod lying on the bank next to him. The other had a canvas creel hanging from his shoulder. They seemed oblivious, or at least unconcerned, with the Hendrickson spinners flitting above the stream, nor with the occasional rising fish in front of them.

Their casts did the job, but had a pragmatic feel to them. There was no artistry here; this was what they did when the fish stopped hitting worms or spinners. And on this evening, they were my brothers. But unfortunately, my brothers were clearly in the only sweet spot that held the majority of the trout. The upstream angler landed and creeled three before a trout finally slashed at the caddis dry I had by now substituted for the usual Woolly Bugger.

I landed the fish, a fat fourteen-inch brown that thankfully sported a malformed pectoral—the telltale sign of a genuine stocker. This automatically made it bound for the larder, and I killed it without a second thought. But it turned out to be the only fish at least in that spot, so I moved upstream to a deeper pool.

Another fly angler was already there, making basic casts as an idle spinning rod stuck out from the top of his waders. A stringer with a lone trout dangled from his belt. I gave this other angling brother a wide berth and fished the head of the pool where a few trout rose in the fading light.

And here is where it happened. I decided to match the hatch—rarely needed on stockers—with a size 14 Hendrickson dry. On the second drift, a trout took then tore off downstream taking line.

The fish was no monster but it fought surprisingly hard, giving me a slightly sinking feeling. And then a minute or so later, my fear was confirmed. It was a beautifully spotted thirteen-inch brown, complete with pectoral fins as wide as a ladies' fan: no eroded tail, no washed out colors, no incomplete gill plates.

Yes, this was a wild trout, born in the Catskills, surviving otters and kingfishers and now rising to mayflies on a warm April evening. And I was . . . disappointed, because I knew that I could never kill such a fish, that it would instead have to be coddled, and worshipped, and gently released like all the other damned Sacred Wild Fish.

Meanwhile my dead stocker looked particularly lonely lying there with an aluminum wire sticking from its mouth. I eventually left the stream oddly unsatisfied yet content. At least I had one fish for my smoker.

The November Striper Diaries

The following three journal entries, logged years apart over nearly two decades of surf fishing for striped bass, are eerily similar, containing the familiar themes all surfcasters grapple with: solitude versus loneliness, confidence versus self-doubt, and triumph versus humiliation. I think I'll go fishing again tonight.

Night of the Giant Striper—November 13, 1993

Sometimes wind, tide, and moon conspire to make certain November nights perfect for catching a very large striped bass. A front passes; brisk northwest winds flush migrating baitfish from the protection of estuaries. Spring tides seethe and swirl pulling water over bars and through cuts and drop-offs. And the darkness of the new moon draws big stripers into the shallows to gorge.

Last night was one of those nights. I stood alone on a beach bundled against the cold casting into a roaring tide rip. Here, a deep channel runs parallel to the beach before coming to a sandy point. A backwater rushes in from the other side forming a one-hundred-yard-long ripline of confused currents and eddies that eventually organize and continue their journey out to sea.

I cast a midnight-black bottle plug—a wooden swimmer that dives and wobbles like a wounded menhaden or sea herring. In the pitch of a moonless night, it may seem counterintuitive to cast for stripers using such a dark lure. But you are not a striper. "Dark night, dark lures," the sharpies say, and they are right. If the moon was full, a yellow plug that catches reflected moonlight would work best.

Fishing theories aside, I can attest to the black bottle's effectiveness. Almost a year ago to the day, casting this exact plug and in this very same rip, I hooked and landed a forty-two-inch bass that weighed nearly thirty pounds. And much larger fish have come from here. A few seasons earlier, rumors of a sixty-pounder raced up and down the beach like a blitzing school of bluefish.

I start at the beginning of the rip making short, quartering casts downtide. Each time the plug splashes down, I jerk the rod making the lure wobble and dig into the current. Its vibrations transmitted downward through the surf rod to my hands while I reel slowly. Unlike bluefish, which bite at anything as long as it's zipped along at a breakneck speed, striped bass—particularly large ones feeding at night—prefer their lures crawled.

I work the rip as a salmon angler might cover a favorite pool—cast, take a few steps and cast again. The spot is deserted, so I am free to fish at my own pace without the fear of tangling another line. Twinkling lights from distant shorelines are my only company.

I eventually fish my way to the very tip of the point then begin lengthening my casts to cover its farthest reaches.

A dark bruise of surging current rises and swirls perhaps a hundred yards away. It marks the end of the rip and is closest to the security of deep water. This is the spot I am saving until the end; the one I know would hold the largest bass. A big striper will keep back and let the smaller, less experienced fish feed in the shallows closer to the beach. It's similar to the way a trophy trout holds at the tailout of a pool casually sipping spent mayflies while the yearlings waste energy slashing and jumping in the riffles.

A headwind makes it difficult to cast too far, so each time the lure touches down, I keep the bail open and let the tide take the plug farther into the current before beginning my retrieve. With each successive cast, I count five seconds longer. With each cast, the lure comes closer to the magical end of the rip.

I fire the lure out one more time with a grunt. Now I wait a full minute before starting the retrieve. Coils of line slip off the reel, through the guides of the rod, and into the dark Atlantic. Finally, I pick up the line, reel a turn, and feel the plug dig into faster current. I have reached the bruise.

The plug creeps along. I clutch the rod, checking it slightly upward. If a fish hits, even at this distance, it will get the full force of an upward strike. Through fingerless gloves, my index finger maintains contact with the braided line. I dig my boots in the wet sand in anticipation.

The strike is jarring. Even though the big bass is one hundred yards away, I feel its force down the rod, through my hands, and into my boots.

I rear back mightily jabbing back several times and setting the treble hooks into . . . nothingness. I reel a quick turn, and jab the rod again just to make sure the fish hasn't somehow grabbed the plug and swum toward me. Nothing. Just the wobbling of the plug against the ebbing tide.

It is my only hit for the entire night.

There. Now you know how I felt. Night of the giant striper my ass.

A Sandy Hook Beating—November 4, 2006

Things looked bleak. Jim had already reported slow fishing the night before with just one swirl—not even at his lure—to show for several hours' effort. Tonight, I fish solo, but am relegated to a B-list-spot since two fishermen have already laid claim to my favorite point. Three's a crowd there, so I wish my fellow anglers a good skunking before I move on.

Now I stand a mile or so away, casting with just the moon and a few wayward sea ducks as company. This particular beach has a series of sandy points that form rips as the tide drops. I work each one of them methodically. But after a thorough carpet-bombing with various plugs and bucktails, my confidence begins to erode with the ebbing tide with not even a bump or swirl to show for my efforts. But before I surrender fully, I will cast again at the first point where I started. I clip on a yellow bucktail tipped with a red pork rind.

I make the cast and start a slow retrieve. But I am admittedly out of The Zone and already thinking of the car, and its heater, and the late-night bowl of Cheerios I will devour when I got home.

The fish doesn't really hit as much as it is just there. Reflexively, I jab the rod a few times to make sure the hook sticks, then get a few turns of line on the reel before the fish runs . . .

And runs . . .

And runs . . .

For a moment, as the rod strains right down to the cork, I think I have foul-hooked something monstrous and unstoppable like a sturgeon. But eventually the run stops and I feel a few ponderous head shakes, making me realize I have actually hung a really nice bass—the kind of fish that if I lose, I will feel like puking, or I will toss my rod into the drink and call it a season. The bass continues grinding out more yards of line. I literally stand my ground digging my boots into the wet sand and thinking how crappy I will feel if the line breaks or the hook pulls.

But they don't.

And eventually, like when the Grinch finds he had the strength of ten Grinches, plus two, I manage to turn the fish's head out of the current and coax it toward me. Then it's just a few yards away thrashing, and before I know it, I have the shock leader in my gloved hand and am leading the fish up on the beach. Then it's over. Twenty-five pounds of striper—a good forty inches and fat—lie in the sand with sea lice scurrying over its silvery body like Lilliputians on Gulliver crying "There's a giant on the beach!"

I sit next to the bass for a few moments staring up at the stars on a November night. This was the first big fish I have caught since the cancer surgery. Now, a year later, I have just gotten my ass kicked again, but this time by a big striper instead of a tumor. I prefer the striper.

Still sitting, I turn the fish around and slide it back into the rip, where it kicks its wide tail a few times and vanishes.

As more of an afterthought, and because this is technically a fishing report, I dutifully note that I dropped another fish a few casts later then landed a ten-pound bass that broke water a few yards in front of me. It thumped a bucktail as soon as I twitched it past its nose. That was the only sign of a feeding fish I saw on a night that had very nearly proved to be fishless.

Escape from a Dark Place—November 12, 2011

I stand on the beach at midnight. The surf is completely and utterly dead. Hours of nothing. This follows another two hours at sunset before taking a break for dinner.

I am staying at my friend Brian's new beach house. He is eager to break in the place by catching something—anything—in the surf. But it is not happening on this Friday. The hard nor'wester had dirtied the surf, and the few working gannets we saw earlier were a good mile off the beach.

So far, I have snagged a skate on a needlefish plug and foul-hooked a gull on a popper during the daylight session. Not good. A group of bait guys off in the distance hauled in a lone short striper at dusk. Based on their wild celebration, this is not a common event for them.

Now it is dark and the beach cold and empty. My newly resurrected vintage Penn 702 "Greenie" is apparently not lubed correctly and has begun to gum up and slow down, making reeling increasingly difficult. I am using a heavy ten-foot surf rod to get some distance. It now feels as limber as an oak. I let it slip through my hands so it is nearly vertical with the butt dragging in the sand. I am far from the poised surf angler anticipating a strike.

I breathe in salty November air. At least the spot where I stand looks good. The rising moon reveals a cut in the outer bar where current escapes seaward in a river of confused whitewater.

But as fishy as the spot looks, cast after cast proves fruitless. I have now sunken into the dark place where all anglers dread going—the place where confidence erodes to nothingness: where you are no longer fishing; you are wasting your time.

I am longing for Brian to emerge from down the beach somewhere and tell me that we should head in and have a nightcap before going to bed. But Brian is not to be found, so I continue to cast. I am comfortable at least; my waders don't leak and the extra layers feel good against the cold. The gummed-up, old Penn's bearings now force me to crank so slowly I hardly reel at all. I feel as if I could almost fall asleep here, standing still on the edge of the surf with the plug just wobbling away in the current.

The hit comes violent and unexpected and joyous; a Gatorade bath after a come-from-behind touchdown as time expires. The rod lurches downward and the tip points immovably just beyond the first wave. I am so completely flabbergasted I do nothing but walk backwards. No setting the hook; no cranking down to gain line. Just a slow backward march up the beach until the waters part and a nine-pound, thirty-inch striper comes flopping onto the sand, glowing in the moonlight like some rare alloy dredged from the bowels of the earth.

I'm laughing out loud now, moved by what can only be divine intervention from the Fishing Gods. I grab the bass, which has already shaken the hook free and find Brian, who is down the beach a ways in the dark place from where I just pole-vaulted. I stand behind him and utter a single word: "YO." He turns around startled then stares at the fish up and down like Charlie Brown gawking at his newly decorated Christmas tree. Then he begins laughing with me in astonishment.

The fish is laid carefully up the beach away from the waves, and we cast some more with newfound confidence. But the Fishing Gods decide enough is enough for me, and unleash the mother of all bird's nests on my poor reel. Eventually the bass is gutted and dragged off the beach and later made into chowder that tastes particularly satisfying.

Notes on April Shad

Each spring, American shad make their way up the Delaware River to spawn, luring me to the dangerous Lambertville Wing Dam—a low-slung concrete structure that juts almost defiantly into turbid, roaring currents. Slick footing and class III rapids downstream add to the excitement. Fall in here, and you are eel food. The spot is as unpredictable as it is hazardous; entire seasons can go by with the river running either too high, too low, too cold, or too muddy to yield a single shad for me. Yet I continue to fish here. Why? Here are two journal entries that I offer as possible explanations.

The Quest for Roe—April 29, 2006

And so it came to pass that on the penultimate day of the Fourth Month of the Two-Thousand-and-Sixth Year of our Lord, I had still not captured/killed/baked/smoked and/or sautéed a female, egg-laden shad (heretofore referred to as a "roe") for the season. So calls were made to nephew Johnny (age twelve) and Jim (much, much older), and alarms were set for 4:25 a.m. Saturday morning. Destination: Lambertville Wing Dam. Mission: Kill one roe; annoy/release all other fish that get in our way.

Over chocolate Entenmann's donuts, Johnny and I discuss our strategy while driving south on Route 206. I tell Johnny that Jim and I are gunning expressly for shad, and therefore will fish exclusively at the end of the two-hundred-yard-long wing dam. Johnny will be relegated to fishing the big—and much safer—pool at the base of the dam, where the currents are far less treacherous. I need to bring you back to your parents alive, I explain.

But I also assure him I am not sending him to the kiddie pool, as the water below the dam can be a veritable fishbowl of warm- and cool-water species. To bolster his confidence, I tell him that Jim and I have caught some twenty-nine species of fish from the wing dam over the years, and his eyes widen. Of course he then demands the list, so I breathe deep, and feeling a little like I had been asked to name all of the state's capitals, spew out the following: largemouth, smallmouth, and striped bass; American and hickory shad; blueback herring and alewife; brook, brown, and rainbow trout (all mysterious stockers from God knows where); white and channel catfish; yellow and brown bullheads; rock bass; white and yellow perch; bluegill, pumpkinseed and green sunfish; muskie (Jim); fallfish; golden shiner; carp; white sucker; American eel; walleye; black crappie; and last and least the lowly and smelly gizzard shad—my least favorite fish.

By the time I have um-ed and er-ed the last of these, we are pulling into the parking lot. Jim's truck is already here. Our policy has always been that whoever gets there first hauls ass to save the sweet spot at the end of the dam. When we reach the spillway, Jim is just where he should be: perched on the end, casting a shad dart into a chute of whitewater.

Johnny is indifferent about things like baked shad and shad roe, so I set him up with a twister jig for walleye, which frequently run at the same time as the shad. I wish him luck, then turn to meet Jim. When I glance over my shoulder one last time, Johnny is already into a fish—a fourteen-inch walleye it turns out—his first ever. I give Johnny a quick thumbs up before heading to the end of the dam. When I reach Jim, he reports he has taken a small walleye on his way out, but no roe. Thus the Quest remains unfulfilled.

I make my first cast and look back at Johnny who is clutching a bent rod. Another walleye is landed and released. I cast off the end of the dam and wait for a shad to grab my dart. It doesn't. But Johnny hooks up again. And again.

Jim and I bombard the big eddy at the end of the dam with shad darts of various colors and sizes. But it only proves frustrating. I take an eight-inch striper on a dart, foul-hook a beast of a fish that manhandles me before snapping off (undoubtedly a two-hundred-pound sturgeon). Then I hook something that I proudly declare to Jim is "definitely a shad." It takes some line, then heads out into the fast water and shakes its head—all pages from the Book of Shad Tricks. The fish makes one last power dive at the base of the dam. When I pump the rod, I brace my eyes for the blinding gleam of chrome that only comes from a sea-bright shad. Instead I see dull gray. Then whiskers. An interloping two-pound channel cat has gobbled up my dart.

Jim and I are eventually lured to the base of the dam to join Johnny, who is releasing yet another walleye (he will wind up catching twenty-three, up to three pounds). Each of us picks a few more before the fishing eventually slows.

We leave and decide to grab breakfast in downtown Lambertville. It's the day of the annual Shad Festival, so we eat quickly trying to escape an onslaught of tourists ready to buy shad earrings and roe-flavored funnel cakes. Before we leave town, we stop at Lewis Island to watch the haul-seine crew. After much rowing and hauling, they bring in five smelly gizzard shad. The crowd lets out a collective groan.

Then I hatch out a plan, one that will involve heading to the one place that immediately comes to mind when you think of big roe shad—you guessed it—downtown Philadelphia. You see, I have been tracking online reports of fine, fine shad fishing at a place called Fairmount Dam on the Schuylkill River, a big lower tributary to the Delaware. Reports of dozens of American shad and scores of hickories make it sound almost too easy. So the Quest will continue southward. Sadly, Johnny and I bid farewell to Jim who had other Quests to fulfill, most having to do with yardwork and napping.

An hour later, we find it—a twenty-foot-tall spillway with a fish ladder on the near end. Sculling crews from various colleges maintain Victorian boathouses above the dam along the far shore giving the place a sort of Ivy League feel. But the populist shad don't care. They are stacked below the dam, swimming frantically in surprisingly clear water. (I have since learned that the Schuylkill—nicknamed the "Sure-kill"—was once so polluted they mined coal from its bottom from all the tailings being dumped in from upstream.)

Along with American shad, I spot pods of hickories (smaller and more tightly packed), schools of herring, gizzard shad (yuck), a couple of very nice smallmouth holding in the shadows of a boulder, and a pair of large lips attached to a fifteen-pound carp. One unfortunate aspect to all of this: You need a Pennsylvania license to fish here, and I have none. So I suck it up and played guide while Johnny-the-minor fishes. His first cast: buck shad. A few casts later Johnny hooks another that jumps like a tarpon then breaks him off. The river is tidal here, so flooding water forces Johnny off his perch. He winds up on a high concrete bulkhead about twenty feet above the river and continues to hook shad. Occasionally, I borrow the rod for a quick, illegal cast. More times than not, I hook up or get a hit. One angler who fishes a banged-up surf-rod and odd-looking plug seems dumbfounded by all of this. "What sort of bait are you using?" he demands. I tell him a shad dart, and he looks into the river like this was a type of baitfish he could liveline.

It's getting late and still no roe, so I grab . . . er, borrow the rod from Johnny for one more cast. I pitch the dart just below the face of the dam and let it sink deep. The take is immediate and the fish jumps high—nice shad. I walk it down the bulkhead to a spot where Johnny can scramble down the bank to land it for me. "I think it's a roe," I announce. Johnny says no, he thinks it's a buck (oh now *he's* the expert). "No Johnny, it's a roe," I assure him in my best grown-up tone. "Now grab it by the gills and haul it up."

When he hands me the fish—about a three-and-a-half-pounder, I run my fingers along its belly, and receive a gentle squirt of milt from its vent (not that there's anything wrong with that). Like the Kinks' Lola, this shad is a man.

I wind up handing off the fish to another fisherman who keeps asking me: "You gonna keep that one?"

This Quest shall continue . . . upstream.

High, Scary Water—April 18, 2011

Jim and I went to the wing dam for a first-light shad sortie on Saturday. The river had gone up since our last visit. The end of the dam two hundred yards away looked particularly fishy in the high water rushing over it. But it also loomed on the dangerous side with rapids and deep pools boiling and swirling all around it.

There was a time I would have pranced out to the end, high water or not. That time has apparently passed. I urged Jim to stay and fish the less adventuresome base of the dam. Maybe we could hook a walleye—the resident glamour fish on the lower Delaware, but in reality a booby prize since they don't fight, and around here taste like river mud—the opposite of the muscular and toothsome shad.

But just a few fishless casts later in this quiet, less interesting water, I could see Jim staring at the end of the dam with forlorn glances. A minute later, he had seen enough and began wading out. I reluctantly followed. I should add that we both wore life jackets, which are mandatory gear at this spot in early April when the river is high and cold and the shad run best. Every few years a careless fisherman, boater, or canoeist drowns in the rapids below us.

Jim made it to the end with no problem; I stopped three quarters of the way out, the river tugging and surging at my waders. "Can't do it!", I announced, before turning around and shuffling back. Jim shrugged, made a cast, and hooked a shad. Then another. When he hooked his third, I miraculously found my wing dam legs and joined him. Once I finally reached Jim taking baby steps, I realized in fact we were quite safe at the slightly higher rise at the dam's end. And the shad were waiting for me, too; a steady pick that waxed and waned as small pods swam by.

The best part about this spot is the way the shad takes, improbable in any circumstance since they are technically not feeding during their spawning migration. But here, it seems even more unlikely. The fish ascend through a chute of very fast water. Though the edges run slightly slower, it all looks too fast to hold shad—especially one that would feel the need to grab a dart.

Not true, as it turns out. With the right cast, with the right dart, on the right day, with the right water level, with the right cloud cover and barometric pressure, during the right time of year—and if the stars are correctly aligned and you've been good to your mom—you cast, pay out just enough line, flip the bail, then point the rod at your shad dart and watch.

Your dart begins to swim through this seemingly too-fast water—you track its course by watching your line. Then you notice that the swimming speed slows slightly as the river channel deepens. You have entered the taking zone. Then it happens. You feel a tap immediately followed by a violent yank that sends your reel's drag spinning wildly and your rod bucking. A shad has taken the dart then turned headlong into an eight-knot current. You have hooked a wing dam shad.

This is the take that makes me grin during dull work meetings; the hit that lulls me off to sleep in the bowels of winter.

There were several takes like this over the next two hours and I relished each of them. There were also more workman-like grabs when we cast into adjacent eddies. All were happily accepted. A nice mix of roe to buck—about 50/50. Two four-pound roes didn't make it back into the river—one was stuffed and baked whole Saturday night; the second is being pickled.

The bonus fish was a solid three-pound walleye that nailed my red-and white dart then fought like a waterlogged branch. Back in the river it went. Lower river walleye taste like silt; lower river shad taste like spring.

My Pet Brookies

The words "school" and "brook trout" normally do not go together, but that's what lived in the little pool below the culvert. More than a dozen brookies—from yearlings to a few honest ten-inchers—had piled into this waist-deep refuge over the course of a summer. While the rest of the stream shrank to seasonal ankle-high flats, the school swam in their Jacuzzi-sized home holding lazily in the smooth tongue of current that flowed in from the culvert pipe.

And for one glorious summer, they were my pets.

The little stream they lived in happened to run past a cabin I had bought ten years ago in upstate New York. I had come to know the stream well. It rises from a pond then tumbles through a narrow hemlock-lined gorge before flowing past the front porch and under a road. It's tiny, less than six feet wide in most stretches, and serves mostly as a spawning tributary and nursery for rainbows that run up the East Branch of the Delaware River.

But it also contains a resident population of brook trout. Most are too small to interest anyone except the most dedicated native trout specialist. Five inches is average; nine inches is a brute. And the stream is private, posted for nearly its entire length, so fishing pressure remains minimal.

Having any fish living in the brook that summer felt like a miracle of sorts. Three years earlier, a five-hundred-year flood decimated the stream washing away many of the trees that once shaded its pools and stabilized its banks. The town road that paralleled it washed completely away in some spots. When the road was eventually repaired and the bulldozers and backhoes finally moved on, they left behind something that resembled the Los Angeles River. Channelized banks devoid of any vegetation now lined much of its length. The stream itself was precisely

Peas Eddy Brook Native Trout

the width of a bulldozer's blade and just inches deep. The water ran milky and lifeless. When I first saw it after the repairs, I thought decades from now I would shock my grandkids by telling them that the drainage ditch in front of Gramps's cabin once contained brook trout.

The next year I began restoring parts of the streambanks by hammering live willow stakes into them and planting shade trees and shrub seedlings. Slowly the streambed began to transform back into its riffle-pool character. By the second year, I spotted my first brook trout since the flood—probably a recruit from its non-channelized headwaters.

By year three, some of the willows were six feet high, most of the insect life had returned, and the culvert pool—now the deepest part of the stream—had filled up with a school of brook trout.

When I first discovered the school of trout living in this tiny spot, they were hard to resist. I would stalk them and cast dry flies with a little two-weight rod. Admittedly, they were not hard to catch as long as you approached the pool from downstream. Any reasonably presented fly that floated would be swiftly attacked as soon as it touched down. Then I would release them—perfect, sleek fish with olive backs, vermillion fins, and blue and white spots.

But after a while, I began thinking back to all they had been through—the flood, the bulldozers, and now me. So I decided to leave them alone. Sort of.

I discovered the art and discipline of fly fishing without setting the hook. I would approach the fish, this time from upstream, and stand adjacent to the culvert. The school would inevitably spook as soon as they saw my shadow, zipping this way and that until they found shelter under some of the pool's larger boulders. But I found if I remained there for a few minutes, some of the trout would begin to ease out of their shelters.

Then I would cast a big meaty attracter fly—a Stimulator or a Turk's Tarantula would usually do the trick. I'd dap it on the surface letting it skip and bounce while holding the rod high.

Sometimes it took a minute or two, but a trout would almost always eventually rise up and take a whack at the fly. And with all the self-control I could muster, I would do nothing: no firm lifting of the rod to set the hook, no snap of the wrist. I would just let the fish spit the fly out. Sometimes, the same trout would come back half a dozen times or others would join in. And I, a rock of self-control, would continue to do nothing. After the trout had caught on and stopped rising, I would reel up my line and leave them alone, overcome with a strange feeling of contentment. When friends visited, I would sometimes show off my fine crop of trout using this trick. But if they asked if they could make a cast or two, I would change the subject and hurry them away. I had grown increasingly protective of my personal school of trout.

Later that same summer, the surrounding woods were infested with a plague of forest tent caterpillars that stripped bare entire lots of hardwoods. The one benefit of these otherwise repugnant creatures occurred after they molted into moths. At night, I would collect a dozen or so that gathered around the porch light and keep them in a jar for the next morning. Then I would wade into the stream above the culvert, shake the jar to stun them, and then sprinkle them in the water where they would float downstream and into the pool.

The stunned moths would then motorboat and skitter along the surface. But not for long. Trout would come up like piranhas slashing and gulping until every last one of the hapless bugs was gone. Watching this display for the first time, I quickly dismissed any thoughts of brook trout as exclusively delicate feeders that daintily sip mayflies. They can be as ferocious as bluefish.

By the end of the summer, the trout had grown larger and I began to contemplate whether I should harvest a few. I could cook breakfast trout—fried in bacon fat. I'd serve them with some fried potatoes and a pot of strong coffee. Yeah, maybe just before the end of the season I would do that. Maybe . . .

August turned into September and I still hadn't brought myself to harvest any of the trout. I may have become a trout hoarder. After all, they were my fish, living in my stream that I had restored. I coddled them, and fed them, and played with them.

Then one day it ended.

It was a warm morning. After breakfast, I walked outside for what had become my weekend routine: taking a peek into the culvert pool to see my fine crop of brook trout.

Except this time when I approached the pool to admire my pets there was something else swimming there—a silver-gray duck with a reddish head. A merganser. For those who don't know, mergansers are extremely agile diving ducks whose favorite food is—you guessed it—trout. They are nicknamed helldivers because that's what they are on trout populations in small streams—pure hell.

The merganser sped downstream like the proverbial fox fleeing the henhouse. But it left a calling card. At the tail of the pool, there was a small, shiny object lying on the bottom. I took a few steps and then saw it clearly—the head and some entrails of one of my brook trout.

I looked around the pool in a state of panic hoping for any sign of the school. I saw nothing. Maybe, I hoped, they were all safely hidden under the boulders. I decided to rest the pool for an hour and take another look. When I did, it was empty. The next day, it remained empty, as it did for the rest of the season. The merganser had eaten the entire school, a stark lesson that one fisherman's coddled pet brook trout were just another's breakfast.

A December Striper Report

It's Saturday afternoon, December 5, 2015, and I'm heading farther and farther southward on the Garden State Parkway. Jim, who has hit the road an hour before me, is providing real-time dispatches: Sandy Hook is dead; Deal is on life support with a few boats anchored offshore and anglers standing around with hands in pockets.

Meanwhile I had heard that on Friday, Seaside Park was ground zero for birds, bass, and bunker, but I know all of angling humanity will now be there. So I take a gamble and go even farther south to the next barrier beach, tony Long Beach Island. There, I am hoping I would find the three Bs, but not have to deal with a mob of anglers since I am with Mimi and Finn—now aged ten—and eager to catch his first striper.

But LBI is all quiet. We stand on a very pretty beach with nice structure and look around. There is even a flock of gulls maybe three quarters of a mile offshore. Meanwhile, Jim's latest dispatch comes in from Seaside: blitzes happening a few casts away STOP. Getting ready to launch kayak STOP.

What to do? It is a beautiful day on a lovely beach with no one around. We can easily spend the rest of the afternoon here and it will be a fine day. Maybe one of us will luck into a schoolie bass just blind casting, or maybe a flounder or hickory shad that I hear are around, too.

But this isn't about a pleasant afternoon; this is about GLORY, a taste of the action I had seen over the past few weeks: bass blasting bunker schools, hundreds of birds diving, mayhem . . . carnage. . . . And, by God, I want to share it with my family. Even if they experience it for mere minutes, it will be far more memorable than just another "nice day on the beach."

So I announce my decision: We are heading into the belly of the beast—to Seaside.

Forty-five minutes of semi-frantic driving later, we arrive to parking lots full of tripped-out four-wheel drives and serious fishermen everywhere. But they, too, are mostly standing around with their hands in their pockets. Half a mile off the beach, gulls are so thick it looks like fog. Two dozen boats and a few kayaks bob among them. A quick call to Jim confirms that he is in the thick of it, and yes it is pandemonium. A school of peanut bunker has literally just jumped into the back of his kayak immediately followed by a keeper striper that lands on top of a hatch, lays there for a few seconds, then flops back over the side. "I gotta go," he says. "Big blitz in front of me." Click. Bastard.

So we pick a spot among the picket fence of surf fishermen and begin casting. I hope for the best, but now know it will probably be one of those days watching the boat guys have a fine time of it, while the great unwashed masses (us) go fishless. Won't be the first time; won't be the last time. A seal cruising in front of me confirms that it will be slim pickings here.

Now it is 4:00 p.m. and the sun is low on the horizon. A quick scan to the south and . . . are those birds close to the beach maybe half a mile away? It certainly looks that way.

"We're moving," I announce. And we do, piling back into the car and making our way south. Just as the road makes a hard turn away from the beach, I can see gulls over the dunes that do in fact look close. So we park and scramble out of the car and along a short trail—me first, followed by Finn and Mimi. I now clearly see gulls diving crazily just beyond the breakers. My God, maybe we will actually catch something.

"Come on Finn, hurry!" I yell. As I run closer, I'm thinking I may have to make a long cast, hook a bass, and let the little guy reel it in. But then a living room–sized school of peanut bunker erupts literally on the beach with bass cartwheeling and crashing everywhere.

"FINN!! RUN!!!!" I sprint ahead as the bass continue to all-out blitz. But just as I approach the school relishing that true angling glory will in fact happen, the Fishing Gods throw me a wicked, wicked curve when my braided line somehow gets caught in a stupidly placed zipper on the underarm of my fleece jacket. I can't get to it and can't cast. Mimi offers to help, but I tell her: "No! fish!" She casts and immediately hooks a bass.

The school is now fifteen feet away.

Finn has caught up and is awaiting instructions. I tell him: "FINN! CAST!" He does but his lure somehow lands way off the mark. With loose braid becoming more and more tangled around me, I make a decision. I will cast for Finn.

"FINN! GIVE ME THE ROD!" He does and I pitch the lure just beyond the breaking school. I hand it back to him. He reels, gets hit, but misses. The next wave sends dozens of peanut bunker flopping all around us on the beach. You can kick the bass they are so close. Now, I am too tangled to hazard another cast for Finn.

So I plead with him: "FINN! CAST AGAIN!" An old-timer who is fishing next to me can sense what's happening. He calls out, too: "Go ahead, boy, cast!"

And Finn does. His lure lands about ten feet to the right of the blitzing fish, but it doesn't matter. His rod lurches and he has hooked his first ever striped bass. He cries out: "I've got one!"

Now I'm jumping around and yelling in Finn's ear: "Walk Backwards!"

But Finn is transfixed as the rod surges seaward a few times. Eventually my words sink in and he slowly backs up the beach. The next wave deposits a nice schoolie onto the sand. Finn has now landed his first striped bass, a fine five-pounder.

A few pictures are taken and the fish is released, but now Finn's line has gotten tangled in my mess. After an incredibly frustrating two minutes with fish continuing to erupt, I tell him to please go get his mother. Mimi later tells me Finn came running up to her and pleaded: "Mom, you've got to help Dad. He's using really bad language."

Mimi untangles me while telling me she just released her third bass.

Finally, unshackled, I make a frantic short cast. On cue, a nice bass rises at the lip of the beach and crushes my popper. It streaks away and I see it is clearly a keeper—maybe ten pounds. As badly as I want to show a big fish to Finn, the Gods won't let me, and the hook pulls. A few casts later, I land a more modest three-pounder that I quickly release.

Meanwhile, Mimi and Finn are now somehow tangled, and I have to intervene with a Leatherman and retie leaders and send each of them back to the surf. By now the school has moved on, but it doesn't matter; they saw it first-hand: Pure, Blinding, White-Hot Glory.

Eventually we catch up with Jim, who reports landing bass after bass and bluefish after bluefish. And then we all wind up at a local tavern a block from the beach. There, over a round of beers, rum, and a large ice water for Finn, plus Jersey Shore–style thin-crust pizza, we retell our stories—all laughing and toasting each other like conquering heroes.

Fish Off

One day last summer I floated a few miles of the Upper Delaware River with my son and two friends. Springtime trout and shad seasons were over and the fishing had become lazy and casual. We drifted in two canoes lobbing spinning lures here and there to rocky shorelines and slow, deep holes. Maybe we would catch a smallmouth or a rock bass.

The sky was a beautiful blue dome that day with the summer sun shining brightly. I had just lathered up with sunscreen when something hit my diving plug. After an unremarkable battle, I landed a foot-long fallfish.

I grabbed it behind the head and began working the hooks free like I had done countless times before.

The rear treble hook was caught in the fish's upper jaw, while the front hook dangled freely. These particular hooks were "chemically sharpened," according to the package. Though I still don't know exactly what that means, I can attest that they were almost dangerously sharp. They would catch on fingers and knuckles whenever I tied the plug on or even reached into my tackle box to grab it.

The fallfish, which up to this point lay limply in my hand, now suddenly began to struggle. As it writhed and wriggled, its slime coat conspired with the greasy sunscreen on my hands to make it feel like a wet bar of soap.

Uh-oh.

One last wiggle and I felt the fish squirt out of my grip followed by the sharp pain of treble hooks burying deeply into flesh. The fallfish continued thrashing about driving the hooks in even deeper.

When the fish stopped, I could see that the front treble had indeed found home. One of its three hooks was buried into the tip of my thumb;

another was deep into the tip of my middle finger fusing them together into a strange O shape. The fallfish still dangled from the rear treble.

My nine-year-old son saw the carnage and yelled: "Oh no, blood!" and looked away in horror.

So I decided to head for the shore to fully assess the damage. But when I reached for my canoe paddle I realized I could no longer hold anything in my encumbered right hand. So I calmly told my son to paddle as best he could. He did, and I managed to grab onto some tall grass with my free hand to keep us from drifting farther downstream.

I called to my friends in the other canoe and told them what happened. One of them yelled back that he had a pair of pliers and began paddling over.

As I waited, I sighed and gazed upward to the Fishing Gods. Again? Really?

I have been fishing obsessively for nearly thirty-five years. I taught myself how to fly fish, how to read a beach, and how to surfcast for stripers at midnight. I can tie my own flies that actually catch trout. I have fished in exotic locations for some of the world's great gamefish.

And this is what it's come too? A shitty fallfish dangling from my fingers now painfully contorted into an OK symbol?

My friend handed me his pliers. Luckily, I had previously de-barbed the treble hooks—wisdom that comes from too many emergency room visits and fishing trips cut short. Once hooked, twice de-barbed is my credo.

The first thing I did was remove the fallfish and toss it back into the river (I was briefly tempted to fling it into the bushes). Then came careful probing of the entry wound as hook number one was eventually eased out of my middle finger. A few minutes later, hook number two slid out of my thumb.

Later that day, when I got home, I soaked my hand in hot water then swabbed it with antibiotic cream and bandaged it up. In a few days it would feel just a little sore.

So was it worth it, this ill-fated canoe trip?

I retold the story the other night to a few fishing buddies over dinner who guffawed at my ridiculous predicament—my hand stuck in an OK symbol, me helplessly drifting downriver and my son not bearing to look.

Of course, this was followed by a flurry of additional "the time I got hooked" stories. I learned about the time John's friend hooked him in the scalp with a huge striper plug—and they both kept fishing ("We just got there," John explained to us matter-of-factly). Then it was Dave's turn, recalling a terrifying story of hooking his brother in the eyelid while fly casting for landlocked salmon on Lake George. Luckily it turned out to be a superficial wound with the fly falling out on their way to the emergency room. But it earned him legendary status among a group of Lake George regulars who would recognize Dave each season and ask: "Are you the guy who hooked your brother in the eyeball?" And then Rob got to retell his harrowing salmon-fly-in-the-neck tale.

Worth it? As we laughed over beers on a warm summer's evening, telling and retelling our hapless stories of lost fish and bad luck, I knew that absolutely and definitely, it damn well was.